ASTRO TURF

ASTRO TURF

The Private Life
of
Rocket Science

M. G. LORD

Walker & Company
New York

First published in the United States of America in 2005 by
Walker Publishing Company, Inc.

Published simultaneously in Canada
by Fitzhenry and Whiteside, Markham, Ontario L3R 4T8

For information about permission to reproduce selections from
this book, write to Permissions, Walker & Company,
104 Fifth Avenue, New York, New York 10011.

Library of Congress Cataloging-in-Publication Data

Lord, M. G.
    Astro turf : the private life of rocket science / by M. G. Lord.
        p.   cm.
    Includes bibliographical references and index.
    ISBN 0-8027-1427-7 (alk. paper)
    1. Rocketry.   2. Jet Propulsion Laboratory (U.S.)—History.
    I. Title.

TL782.L67   2005
629.4'072'079493—dc22

                                                            2004057158

This book was prepared with the generous support
of the Alfred P. Sloan Foundation.

Art Credits: Pages xii, 3, 37, 157, and 205 photographs by Patricia Williams. Pages 7, 46, 56, 61, 73, 87, 91, 113, 122, 125, 137, 142, 149, 174, 198, 202, 217, and 218 courtesy of NASA/JPL-Caltech. Pages 22, 24, 27, 29, 34, 128, and 211 author's collection. Page 54 Charles Kohlhase. Page 75 courtesy of NASA/GRIN Collection. Pages 77 and 82 courtesy of NASA/NIX Collection. Page 105 *Leonardo*. Page 118 Caryn R. Leland. Page 160 courtesy of Boeing. Page 169 courtesy of Hewlett-Packard. Page 189 Science Photo Library, London. Page 216 Gene Blevins/*L.A. Daily News*/CORBIS.

Book design by Maura Fadden Rosenthal/Mspace*ny*

Visit Walker & Company's Web site at www.walkerbooks.com

Printed in the United States of America

2   4   6   8   10   9   7   5   3   1

To Robin, Nick, Zoe, and Maya

Never send a man to do a robot's job.

— *Gentry Lee*

✳ ✳ ✳

Don't marry an engineer, Dolores, marry an artist.

They have more home life.

— *Robert A. Heinlein*

CONTENTS

ASTRO TURF

Introduction

or

THE BRIDE COMES TO YELLOW SKY

This was not what I had expected.

On a hot afternoon in March 1997, I rolled into the parking lot of the Jet Propulsion Laboratory in Pasadena, California, where the speed limit, in step with scientific convention, was posted in kilometers per hour. Thick brown air hung over the San Gabriel Mountains, veiling the lab—a 177-acre patch of sixties-style buildings scattered in the foothills—in eye-stinging murk. I walked though the guard gate and headed east on Mariner Road, named for a series of robotic spacecraft that flew by Mars and Venus in the 1960s. Then I turned north up a steep hill on Surveyor Road, named for a succession of robots that preceded Neil Armstrong to the moon.

Since the National Aeronautics and Space Administration was formed in 1958, JPL, as the lab is abbreviated, has been a regional NASA center, of which there are fourteen. It is managed, however, by the nearby California Institute of Technology. This may seem a slight consideration, but in terms of culture, it is critical.

Opposite: The Sergeant and Corporal missiles rise from a display stand near concrete cooling towers at JPL.

Beginning with Explorer 1, the first U.S. satellite, launched in 1958, JPL has had a hand in the design or flight of nearly every major U.S. planetary mission. Although it puts on a snazzy show when its robots land on or travel close to other planets, it doesn't throw open its doors to outsiders. You can't get "on lab," as the natives say, without a photo identification badge. Even if you could, you'd be hard-pressed to find your way around. Buildings don't have names, but rather numbers, and addresses reflect the sequence in which the structures were built, not what is next door. Building 264, for example, is located between Buildings 167 and 198.

The grounds were silent, except for ominous mechanical sounds—a hiss from a liquid nitrogen tank, the clatter of droplets from a concrete cooling tower. Two immense penile structures stood at the intersection of Sergeant Road: the Corporal and Sergeant missiles. The former, visitors can learn from a helpful sign, was "America's first operational surface-to-surface missile," designed by JPL in the 1940s when the lab was run by the U.S. Army. The latter, developed a few years later, was a more evolved weapon; it could deliver a conventional or nuclear warhead seventy-five to eighty miles and featured "an inertial guidance system invulnerable to electronic countermeasures."

Up ahead, in the Mars Yard, a model of the plucky Sojourner rover, a brainy shoe box on wheels, scuttled over rust red stones for the benefit of a television crew. In the spring of 1997, Sojourner had not yet become a starlet. Three months hence, "she," as the engineers say, would distinguish herself on the Mars Pathfinder mission.

I headed for a trailer at the crest of the hill. There Donna Shirley, director of the Mars Exploration program, and the

A model of the Sojourner Rover, which rolled around on Mars in 1997.

highest-ranking woman on the lab's technical side, was holding the final class in her four-day seminar "Managing Creativity." Like Sojourner, Shirley would go on to be a household figure during the Pathfinder landing, holding forth in a boxy Mars red suit as a commentator on Cable News Network. Leading her class, Shirley wore jeans, Reeboks, and a tailored man's shirt, suited to clambering over fake Mars dust with her mechanical protégée.

Shirley's coinstructor was Alice Fairhurst, a defiant teenager in the body of a woman born in 1935. Two years earlier, with her daughter-in-law, Fairhurst had published a book on Meyers-

Briggs personality types, with which the class had been work-ing. Each of us, at our deepest core, we learned, harbors an ani-mal archetype. Although "foxes," as nonreflective, quick-to-act, professional-athlete types are known, were sparse in our group, we had a solid representation of utopian "unicorns," bustling "beavers," and, not surprisingly, ultracerebral "owls." Fairhurst, you knew, could tell if you were a beaver or an owl from, say, the pressure of your handshake. But she never imposed an iden-tity. She let you discover it for yourself.

The lights were dimmed and the venetian blinds drawn. We filed in silently and, as instructed, removed our shoes. The inti-macy was almost unbearable. My face burned with embarrass-ment at the sight of a blue-white toe peeking through a tear in the brown sock of a male engineer across from me. I sat with Hector Del Castillo and Christopher Hartsough, owls that I had befriended in the course. Del Castillo is a physicist who special-izes in optics; Hartsough, a software engineer. I dug my feet into the hard blue industrial carpet.

Shirley flicked a switch on a portable boom box, liberating pings of New Age music. A swooshing brought to mind the Pa-cific Ocean, pounding twenty miles away. "Let your eyes relax and your lips relax and your cheeks relax," she intoned. "Relax your face, your mouth, and the very tip of your tongue." I heard the rustle of breath. "Relax your legs, your feet, your knees, your toes." A low whine arose from the speakers, like the cry of a whale.

If we relaxed and released and regressed, we would reach the target of our meditation and the wellspring of our creativity: our inner child.

I tried to imagine my father, an aerospace engineer who had died in 1994, in the room, but I could not. This was not how he had described JPL in the 1960s.

"You may need to overcome deep-seated, forgotten fears from childhood experiences," Shirley continued. "You may need to do things that scare you to death to prove you are no longer a frightened child."

I was torn between being a reporter or a participant— watching the engineers and physicists squirm or shutting my eyes. But Shirley was persuasive.

"I want you to remember an important, positive childhood experience that formed a belief system that you have," Shirley said. "Try to remember as far back as you can."

My lids grew heavy. An orange and purple afterimage of the Sergeant and Corporal missiles melted into the silver pen-and-pencil set on my father's desk.

"Where was the event?" she went on. "Who was there?"

I was in La Jolla, California, on the floor of my room, from which you could see the desk in my dad's study. I remembered two smells—both beguiling to me, though not necessarily to others. They were melting plastic and skunk. The latter emerged from the decrepit foam rubber lining of a pressure helmet that my father had dug out of the trash at General Dynamics-Convair, where he worked in the early 1960s, and presented to me. However stinky and disgusting, the helmet was authentic. It trailed two hoses, and its visor could be sealed shut. The object elevated my stature among my fellow fourth-graders. I wore it for Halloween, class presentations, entertaining after-school guests. At night I slept with it.

The melting smell came from my Mattel Vacuform—a toy that heated a piece of plastic, the size of a slice of American cheese, until it softened and could be molded into an object, like a model airplane. That day, I had made a dozen or so generic fighter jets. My father, amused by this, summoned my mother into the hall to watch. I pretended not to notice.

"How did the experience make you feel?" Shirley continued.

Accomplished. Interested in engineering.

"Now write down an event as an adult where your behavior was influenced by this experience."

In response to the question, I stole a look at the room. "Engineers scribbling," I wrote. "Take orders well. No evident smirks." The experience had left its mark. I had come to JPL to investigate the culture of engineering. My father's death had not shut the door on our relationship; it had exposed doors I dared not open in his lifetime. I wanted to understand who he was and what had made him that way.

Next Shirley asked us to dredge up a darker memory—one that involved pain or a sense of being thwarted.

Again I thought of my father, and the period in the middle 1960s during which he vanished. He was not abducted by aliens or relocated to the other side of the world. What kept him away was Mariner Mars 69, a mission to send two robotic spacecraft to Mars, on which he had been hired as a contractor. My father worked for Northrop Corporation, and JPL, which managed the mission, had subcontracted with this firm to modify the "bus," or body, of the spacecraft.

The late 1960s were not a happy time in our family of three. My mother, barely fifty, was dying of cancer, her frightened,

One of the two Mariner 69 spacecraft that flew by Mars in the summer of 1969.

five-foot-ten-inch body wasted to about one hundred pounds. I was pretty small and frightened myself, shuttling between the unique hell that was home and the garden-variety hell of junior high. What we needed was a full-time husband and father. What we had was a cold-war-era rocket engineer, who embraced the values of his profession: work over family, masculine over feminine, repression over emotion. Whatever grief he may have carried, he remained a silent, archetypal midcentury male.

Oddly, at the same time that I resented his absence, I was fascinated by what he did. The probes he worked on were scouts, sending home thrilling glimpses of unexplored worlds. All were about hope, expansion, the future, none of which, at my mother's deathbed, would otherwise have crossed my mind. They had an almost mystical significance. The few moments of intimacy I shared with him were when he explained them to me.

Tony Spear, the project manager on Pathfinder, once described engineering as "the study of failure." You find out why something didn't work so you can make it work the next time. Pathfinder was hard for its creators to deal with because each component actually did its job. "Failure," however, was an apt term to describe my relationship with my father.

We fended for ourselves after my mother died—a housekeeper was ruled out by daunting unpaid medical bills—and I am not entirely sure how we got through. Neither he nor I could cook, and he refused to try, viewing the process as an affront to his masculine dignity. At fourteen, a sophomore in high school, I was not exactly Julia Child, and I grew to hate food preparation, baffled that anyone could confuse the excruciating process with love. Laundry was another ordeal. Never mind that my father had designed flight controls for fighter jets and spacecraft. He would not understand the cycles on the washing machine. Such knowledge, it seemed, was precluded by his gender. It was useless to explain. When his heaps of socks and underwear became too revolting, I scooped them off the floor—the hamper, too, was a challenge—and cleaned them.

The quest for dinner took us to grim "family" restaurants or poorly lit cocktail lounges. Our favorite was a cheesy place on the top floor of the Holiday Inn overlooking the airport in Long Beach, California, where we had moved from La Jolla. The food was awful—limp, breaded halibut, tasteless iceberg lettuce, freakishly emerald peas. But we loved the place. If we pretended to watch the planes, we didn't have to talk to each other.

My father did not push me to follow in his footsteps. He grudgingly helped with science or math homework, shoving aside his second martini to disparage some shortcoming in my mathematical logic. He wanted me to hone my cooking and cleaning skills, perhaps to spare him the chore of finding a new wife. But I was driven by terror. If I did not do well in rock-hard, number-filled courses, for which, heredity notwithstanding, I had no aptitude, I would not get into college. I would be stuck in Long Beach, California, mothering him, for the rest of my life.

Perhaps it was because he was born in 1906, and his only child didn't come along until his fifties. Perhaps it was because he had worked in an environment that, except for a few secretaries, was exclusively male. Perhaps it was because his late wife had chucked teaching high-school chemistry and, later, working for a public utility, to keep his house. But my father had some very weird ideas about women. He couldn't have been more supportive of my intellectual adventures when I was an androgynous kid.

With the advent of puberty, however, that changed. Never mind that I was as much of a tomboy as a gangling, unathletic girl could be. I had become a woman. Biology was destiny. It was time to forget my brain. I had been born into the servant-gender. He expected me to serve.

The memory hurt.

I glanced around the room, amazed and shamed. My inner cynic, I had assumed, was invulnerable to this inner-child stuff. Twenty years ago, when it was fashionable, I had derided it in print. Yet here, today, this fusty exercise had provoked genuine emotion—in people notorious for their detachment. Tears streamed down Fairhurst's face. Men winced.

Shirley pressed on. Recalling these memories was not enough. We had to heal them, through "inner bonding," a technique she had learned from Dr. Margie Paul, a therapist and the author of *Healing Your Aloneness*. Shirley urged us to dramatize a dialogue—or, as she put it, "to dialogue"—with our inner children about these experiences. But don't expect it to go smoothly: "Your inner child may be mad at you."

The technique worked best, she assured us, if we chose an object, like a doll or toy, to embody our inner children. She held up a red-and-green stuffed clown by the scruff of its neck. This object, she informed us, would stand for her inner child. She would enact the part of parent.

I looked around again, checking to see that the faces matched the photos on the ID badges. It was hard to believe this was happening in the birthplace of the Sergeant and Corporal missiles, the base of operations for every significant planetary mission of the last thirty-five years.

"How did you feel when Dad went off to the navy, Donna?" she asked the stuffed clown. Her father had left home when she was three to serve in World War II, first as a ship's doctor in the Pacific, then during the occupation of Japan.

"I felt really lonely," she answered in a higher-pitched, squeaky voice. "I remember riding my horse. We were in Oklahoma in the front yard."

She retrieved more fragments of memory, which she wove into a perception about herself. This early trauma affected how she related to her bosses, coworkers, subordinates, and family. It even explained why she made jokes all the time. "I would try to get people to like me because I didn't want them to leave," she said.

Tough perceptions like these, she told us, were the gifts of inner bonding. As "a hardheaded, practical engineer," she had found it tough to hear her "inner conversations, much less write them down."

People at JPL, she acknowledged, are used to high achievement, to doing everything perfectly the first time. But this exercise opened up a whole other side of your brain—the right side. It's difficult to change your approach to the world when that approach seems to work. Yet gaining access to your creative side, she assured, would be worth the struggle.

This seemed to put Mike, a planetary scientist, at ease. The meditation, he revealed, had "felt strange—and was hard to do in front of people."

Del Castillo thought he might have gotten more out of the exercise had it not taken place in a dark room after lunch. "I just about fell asleep," he admitted.

"You know," said Dawn, a prickly technician who had been goading Shirley all week, "a lot of people don't believe there is an inner child."

"It's a *metaphor*," Shirley rejoined. "A metaphor! Think of your inner child as the sum of your childhood experiences. It's a way of viewing events in your past that influence your present behavior."

Unsuccessful at masking her irritation, Shirley called a recess. I walked outside and filled my lungs with the gritty air. The smog would no doubt shave years off my life, but I was pleased to be inside it. In 1971, Mariner 9 reached Martian orbit and was supposed to have immediately taken pictures of the planet's surface. But it couldn't see very much. A planetwide dust storm came between it and its goal. Sometimes you have to be in the murk or under the murk to see what you came to see.

∗   ∗   ∗

Mariner 9 was a success because it had not been rigidly programmed, unlike the two Soviet Mars probes that had also launched in 1971. Unable to adjust to the Martian weather, the Soviet ships, each of which contained a lander and an orbiter, followed preprogrammed orders. The landers plunged to the Martian surface, where, blinded, they perished. But JPL engineers sent fresh commands to Mariner 9. It adapted to the storm, beginning its photography in January 1972, when the winds had calmed and the dust had settled. Mariner 9 impressed me with the need to be flexible. In 1997, when I first approached JPL, I had some rigid ideas of what I would find—based on childhood memories and the myth of the so-called rocket scientist in popular culture. But ever since I arrived, my preconceptions were belied by what I actually found. Whether I liked it or not, I would have to reprogram as I went along.

Metaphors were important to Donna Shirley. She had spent the first day of class teaching us how to use figurative language. Thinking in metaphors, she believed, was a way to get a handle on abstract concepts. If a scientist or engineer couldn't construct a metaphor for an abstraction, chances were he or she didn't understand it. Shirley had no patience with NASA jargon, shorthand, or catchword acronyms. To "think out of the box," as she put it, you had to find original ways to express your ideas.

Yet Shirley had not evolved into an engineer-poet overnight. The transformation began in 1984, she revealed, when she underwent training in est—an experience that she described, in what can scarcely be called selling terms, as "a potpourri of techniques adapted from brainwashing, Alcoholics Anonymous, and Zen Buddhism." Its goal was "to shock people out of their conventional belief systems."

I made a snide remark about est to Hartsough but was sternly reproved. He, too, was an est alumnus, as were many people at JPL. He reminded me that Richard Feynman, the legendary Caltech physicist who began his career on the Manhattan Project and publicly explained the cause of the 1986 space shuttle Challenger explosion, had been a close buddy of est founder Werner Erhard.

Humbled, I shut up. The class moved on to a fresh topic: "finding your passion," Shirley's version of "follow your bliss."

"So what excites you?" she asked a male class member.

"I love to balance my checkbook," he said. "I do it for recreation."

Shirley pretended to be stunned. "Will you marry me?" she joked.

Although the class would not meet again, Shirley urged us to locate our passions and draft a plan to realize them. She had plotted several such trajectories for herself, and the exercise had helped keep her career on track.

"Are you going to make this a mandatory course for managers?" Del Castillo wanted to know.

Shirley wouldn't commit herself, conserving her energy, I suspected, for other, more critical battles.

✳   ✳   ✳

On July 4, 1997, I was backstage for Pathfinder's triumphant touchdown, as well as for the heartbreaking press conference in December 1999, when the Mars Polar Lander and Deep Space 2 spacecraft were revealed to be lost. But these events, despite their extremes of emotion, seemed unexceptional. Joy is joy; pain is pain; and the lab had seen wins and losses before. Shirley's seminar, however, was too weird to fit in an existing category. It elbowed its way to the front of my consciousness. In some ways, it seemed as shocking as the loss of two missions—a wacky emblem of a sea change in the laboratory's culture.

"Space," in the words of the legendary *Star Trek* voice-over, can be viewed as "the final frontier," and in the 1960s, this meant a wilderness of gruff men swashbuckling their way around Mission Control. Shirley's course, however, was a metaphor for the transformation of that frontier, even as Stephen Crane's short story "The Bride Comes to Yellow Sky" was a fictive marker for the domestication of the American West. In Crane's tale, published in 1898, a small-town sheriff returns with his new wife to Yellow Sky, Texas, where the lawman's archenemy, a notorious gunslinger, has downed a bottle of whiskey and is spoiling for a fight. When "Scratchy," the gunman, sees the bride, however, he backs off. Marriage—indeed, the mere presence of a woman—is a "foreign condition." Flummoxed, frozen, he is "a creature allowed a glimpse of another world."

To be sure, Shirley knew how to handle gruff men. In meetings I watched her banter, throw around acronyms, and, on days when she had to crawl over Mars rocks, dress like one of the guys. Yet she also paved the way for "another world," or what Scratchy, one hundred years hence and toting a slide rule instead of a shotgun, would have viewed as such. In the 1970s, she lobbied for things that were then controversial, like the day-care center that today's two-career couples take for granted. Its founding, less than a decade after the end of the Vietnam War, was an unlikely collaboration between a long-haired, self-proclaimed hippie and the lab's deputy director, a retired three-star general. Of course Shirley was not the only woman pioneer. Nor did she have a hand in every milestone. Many changes in the laboratory culture simply reflected changes in the larger culture. But a great chasm was crossed in the years between 1959, when most women on lab answered phones or operated mechanical counting machines, and 1999, when the project manager, chief scientist, and lead engineer on the Deep Space 2 mission all were women.

Shirley's management course showed me a way to organize my discoveries. During its first three days, she took concepts from chaos theory and applied them to management problems. Fractals, for instance, suggest ways to configure teams within an organization. A fractal pattern involves structures that are self-similar in scale, the way, say, a fern is made up of a spine with branching fronds which are, in turn, made up of a spine with branching leaves. With natural fractals, as opposed to mathematical fractals, it's not the exact shape that's repeated at different levels, but the characteristics of the shape. With a little metaphorical license, the stories of individual engineers or scientists, staff members or contractors, can be looked at as fractals. They relate in a miniature way to the larger history of JPL, which in turn relates to the history of the aerospace industry and of planetary exploration.

What interests me—and what these stories will explore—is the disparity between the stereotype of the cold-war-era rocket engineer, and the men (and more recently, women) who have actually built spacecraft. In many ways, the midcentury so-called rocket scientist can be viewed as an archetype of masculinity, just as, for example, the Barbie doll can be seen as an icon of femininity. Both are abstract ideals, which no human person can ever fully embody. The buzz-cut cowboys of Mission Control, homogeneous as a Rockette kick-line, were a cold-war fiction, along the lines of other cold-war fictions—the notion, for instance, that hard-drinking, womanizing test pilots, when selected to be astronauts, metamorphosed into temperate family men. Many of the engineers who looked, acted, and lived the part later rebelled against it or acknowledged that it had taken a personal toll.

* * *

The cold-war engineering stereotype can be looked at as a "performed identity," something that is socially constructed, not dictated by nature. (In the 1980s, some gender theorists used this term to refer to gender itself, which they considered a matter of approximation and drag.) Engineers, of course, tend not to be self-conscious; they are likely to describe themselves as practical, focused on problem solving, not on conforming to some made-up ideal. Yet in the 1950s and 1960s, engineering was strikingly conformist, even by the standards of a conformist time. Before the federal government mandated diversity, women engineers and engineers of color were far from well represented. And other minorities were explicitly shut out. In 1953, President Dwight Eisenhower issued an executive order that precluded homosexuals from working in federal jobs. Many of the "out" gay engineers and scientists at JPL who lobbied for domestic partner benefits in the 1990s began work during the cold war. They may have been invisible. But they were present.

Nor would every pre–World War II rocket engineer have passed muster in the postwar world. In the 1930s, rocketry was the domain of kooks, nuts, wackos, and science-fiction writers. JPL started out in the 1930s as the Guggenheim Aeronautical Laboratory at the California Institute of Technology (GALCIT). When it was renamed by the army in 1945, however, it was not called the Rocket Propulsion Laboratory, despite the fact that its founders worked with rockets. The public would not have taken the word seriously.

Similarly, two of JPL's first experimenters were not likely to have been welcomed at a midcentury NASA center. One, John Parsons, a self-taught chemist who died in an explosion at age thirty-seven, was less well known for his rocket-fuel formulas than for serving as a priest in the Ordo Templi Orientis, a sex-based (and some say Satanic) cult founded by *Diary of a Drug*

Fiend author Aleister Crowley. And although another JPL pioneer, Frank J. Malina, a Caltech Ph.D. and the son of Czech immigrants who settled in a small Texas town, would seem to have embodied the bootstrap ethos of the American Dream, he, too, is underrecognized. Like many intellectuals during the Depression, Malina flirted with communism, a dalliance that haunted him during the McCarthy era when anti-Communists tried to scrub his accomplishments from the historical record.

This removal seems especially unjust when viewed alongside recently declassified files related to Project Paperclip, a U.S. Army program that permitted certain valuable Nazi scientists to work in this country, despite their compromising war records. At the same time that the FBI struggled to dig up dirt on left-leaning American scientists, the government suppressed information about war crimes committed by Nazis in the U.S. space program. But by the 1980s, after the space race had been won, the disgraces inevitably leaked. In 1984, Arthur Rudolph, head of the U.S. program to develop the Saturn V rocket, fled to Germany, rather than face a denaturalization hearing related to his war crimes. (Rudoph had been in charge of slave labor at the Mittelbau Dora concentration camp, where V-2 missiles were built, and where, near the end of the war, about 150 people died each day.) This was a big, splashy scandal. It brought down a high-ranking former Nazi, second in reputation only to Wernher von Braun, who had died of cancer in 1977.

Restoring luster to a tarnished figure is less splashy. Malina, however, led a fascinating untold life—having fled Pasadena for Paris, France, where he worked first for the United Nations Educational Scientific and Cultural Organization (UNESCO), then as a visual artist, making kinetic sculptures. He also founded

Leonardo, a magazine dedicated to bridging the gulf between science and the arts.

In technical communities, thinking has historically been valued over feeling—a preference for which there is a current, fashionable explanation. During the last ten years, some psychologists and neurobiologists have come to believe that many engineers, physicists, mathematicians, and computer scientists are afflicted with Asperger's syndrome, a mild form of autism named for Hans Asperger, the Viennese pediatrician who cataloged its symptoms in 1944. On the bright side, this syndrome imbues its sufferers with a capacity for intense concentration. But it leaves them at a severe social disadvantage. They are incapable, for instance, of interpreting other people's feelings from their facial expressions, even when those expressions are grossly exaggerated. The disorder has a genetic component. Most boys—and the majority of sufferers are boys—have at least one engineer or computer scientist as a parent.

Thanks to Simon Baron-Cohen, a clinical psychologist and autism researcher at Cambridge University, Asperger's syndrome has acquired a certain cachet. Baron-Cohen examined Sir Isaac Newton (who lived as a hermit without a lover or parents) and Albert Einstein (who was a loner as a boy and tended to repeat sentences) and determined that they suffered from the syndrome. Not all psychologists, however, embrace this posthumous diagnosis. Geniuses, they argue, can be socially inept (and impatient with slower intellects) without being afflicted with a clinical disorder.

Even proponents of the theory maintain that only about 10 percent of engineers, physicists, mathematicians, and computer scientists meet the criteria for a diagnosis. Yet a hyperrational ethos that rewards insensitivity is characteristic of these disciplines. This suggests that a large number of people whose ability to interact with others is not genetically impaired deliberately act as if it were.

I can't evaluate the validity of these posthumous diagnoses; I merely note that they occurred. But the idea of healthy people emulating unhealthy ones made me think of my father, who, for better or worse, has been a constant presence in my investigations. Never mind the differences in age, ethnicity, and background, every engineer I spoke to is, in a psychological sense, a stand-in for him.

This is not to say that he would have felt comfortable with them. Many people whose stories I tell—and whom I have come to admire—do not admire one another. To be sure, they share certain traits: intelligence, dedication to work, and a capacity to see projects through to completion. But their bedrock principles are sometimes in conflict. I don't consider this a problem. It reinforces the idea that the twenty-first-century space program is not monolithic, and what looked like a monolith during the cold war was, in fact, covertly varied. There is, I suppose, one monolith legitimately associated with mid-twentieth-century space exploration—the big black slab that hovered around Keir Dullea in Stanley Kubrick's film *2001: A Space Odyssey*. But that, as the story goes, was not made by people. It was the handiwork of aliens.

Mariner Mars 69

or

A FOOT SOLDIER'S STORY

Shortly before my father died, he became obsessed with being buried next to his father. This, at first, irritated me. He had deposited my mother in a gruesome theme-park cemetery in San Diego, California. And now he wanted to spend eternity in his family's tasteful plot in Hingham, Massachusetts. Particularly because he did not remarry after my mother's death, I was shocked by this apparent abandonment of her. And what felt like an abandonment of me. But believing it unwise to thwart the wishes of the dying, I made the arrangements. Only later would I learn what drove this strange request.

Charles Carroll Lord was a foot soldier, not a general, in the battle for what cold-war strategists called the "high frontier." He was fifty-nine years old when he began work on Mariner Mars 69, an older man in a young man's field. His earlier experiences, however, provide a context for the mission; he had witnessed the way technology was viewed at the dawn of the twentieth century and how it came to be viewed near its end. He also fought in the war of public relations. One of four hundred thousand dads responsible for small components of the space program, he ensured that his near-and-dear had a stake in that program. Or, as Donald Reilly put it in a cold-war-era *New Yorker* cartoon:

"Never forget, Son, that your father sold office supplies to the company that made the box that carried the rocks back from the moon."

Although he worked in a can-do field, my father was not a can-do guy. He never told me, "Don't try" or "Don't dream." He said, "Prepare for disappointment." And when disappointment didn't come, he said, "Wait." This is why he needed to be part of a can-do team. He drew strength from the optimism of the engineers around him.

Oddly, he never wanted to lead the team, just to belong. His reasons were both personal and cultural; and his story, like all stories, is both specific and universal. It begins with his father, Charles Edward Lord, an engineer who received his bachelor of science from the Massachusetts Institute of Technology in 1898.

Unlike my father, my grandfather was a leader. In 1912, after graduating from Georgetown University Law School and serving as a patent attorney at General Electric, he was appointed to head the patent department at International Harvester, a firm whose tractors symbolize the nineteenth century as much as rockets symbolize the twentieth. He was also an instructor at MIT, an associate editor of the *Encyclopedia of Engineering*, and a lecturer in patent law at Marquette University. To hear my father tell it, his father never made a false move. Except one. He was killed in 1919 when a train struck his car. He was forty-three; my father, thirteen.

My father never spoke about my grandfather's death. You'd think it would have surfaced after my mother died, since both of us had been robbed of a parent. But only while babbling on morphine near the end of his life did he reveal his anguish.

My grandfather (Charles Edward Lord), my grandmother (Mary Grace Carroll Lord), and my father (Charles Carroll Lord) in 1906.

I learned details of the accident from the October 19, 1919, issue of the *Harvester World*, an International Harvester in-house newsletter. I found it wedged in one of his overstuffed file cabinets, along with sixteen years of completed crossword puzzles, a cookie tin of rocklike erasers, matchbooks from nightspots that had folded years before I was born, and ten identical clippings of news stories on the suicide by drowning of Monsignor John Storm, a former president of the University of San Diego, who had resigned from the Roman Catholic Church while assigned to our La Jolla parish. (A liberal, Storm and my father had argued often over Vatican II—squabbles that, the clippings suggested, had etched themselves in my father's mind.)

The *Harvester World* evoked a lost age, an age of innocent trust in technology. It fell open to this headline: "Tractors and Buffaloes Assist in Balkan Reconstruction." On the heels of World War I, the American Red Cross had shipped a half-million dollars of plows, reapers, and tractors to what it called "the little Balkan state" of Serbia. Technology would unmake the mess of war, feeding the hungry and hastening the creation of a better life. In the tender photos of farm equipment, I saw the stirrings of America's love affair with the machine, its belief that know-how was the ticket up from ignorance and oppression. "Better farming is the foundation-stone of national progress," asserted a letter to the editor. "Every burden that your machinery lifts from the backs of individual farmers is in precisely the same measure a godsend to humanity and a step forward in national progress."

This was a far cry from the way technology would be regarded in the 1970s, when its most potent legacy was devastation. The atomic bomb had erased two Japanese cities and fused together acres of New Mexico sand. The hydrogen bomb could do much more damage in a single blast. In popular culture, the archetypal physicist of the 1970s was Edward Teller, an architect of the hydrogen bomb, a real-life Dr. Strangelove whose heavy, menacing brows terrified children. Not until 1988 did the popularity of *A Brief History of Time* transform the boyish countenance of Stephen Hawking into the public face of physics.

My father learned well before the 1970s that technology could turn on you—that a machine designed to make your life easier could rip that life apart. According to the magazine, my grandfather had motored to a Harvester plant outside Chicago with an inventor and a machinist. At a railroad crossing, the men noticed a switch engine backing toward them. The inventor hit the brakes, which froze, stranding the car on the tracks. Panicked, the inventor and machinist jumped to safety. My grandfather, however, was trapped. He pushed and clawed his door. When the locomotive hit, he was flung onto the rails, where the engine dragged his body until it stopped. He died, the article said, "after a night of intense suffering, through which he was acutely conscious of his condition, yet courageous and hopeful to the last.

"He leaves a sense of personal loss that is deep and will be lasting," the article continued. But this grief pales next to that of his "widow and his three children, whose loss is even sharper."

Time did not blunt the grief. In ten years, his widow went through all his money, no thanks to her negligent financial advisers and fondness for alcohol. The year was 1928. My father was in college, studying mechanical engineering and studio art, but he had to quit to support the family. When I was a teenager, I remember how, after a few beers, he'd tell strangers that he had been "a pulling guard" for football legend Red Grange at the University of Illinois. He never talked about how he'd left.

"We had our family crash before the crash," explained Uncle Jim, my father's younger brother. We were driving around Wilmette, a Chicago suburb, in his battered plumbing contractor's van, two years after my father died. He was relaxed about the

My father with his sister, Catherine, and brother, James, at Wessagusset Beach near Hingham, Massachusetts, circa 1930.

family's fall from solvency, having never known it. He was three when my grandfather was killed, and, unlike my father, didn't try to follow him into engineering. In front of a vast Georgian revival house—a house that telegraphed second-tier robber baron—he stopped the van. This, he said, was where my father had grown up. I felt queasy. The immensity hit me. I understood my father's defeatism. He had never regained what he had lost.

Although the Lord family went bust in 1928, by the 1930s, the rest of the country had caught up. Jobs were scarce. My father was lucky to have one at Harvester, making mechanical drawings and designing tools. But he felt trapped. It wasn't that he disliked engineering; farm equipment left him cold. As the thunderheads of war darkened over Europe, however, his horizons brightened. The demand for fighter aircraft, and, with it, for mechanical engineers, took off. In 1941, he accepted a job at Consolidated Vultee in San Diego, California. By 1943, he was promoted to supervisor in its New Orleans plant. There, on the shores of Lake Ponchartrain, the firm, whose name would be shortened to Convair, built its proud, propeller-driven flying boat, the PBY.

At midcentury, the archetypal engineer married young and had a bustling family to neglect in favor of his all-consuming work. My father ignored this aspect of the paradigm. He lived with his mother in New Orleans. I never knew her, but I learned more about her than I wanted from the humid letters she sent him when she visited her other children in Chicago. Addressed to her son, a man in his forties, they begin: "My Dear Boy."

Something changed in 1950 when he met my mother, Mary Pfister. He told me she made him laugh. When painful things happened, she turned them into funny stories. Of course he asked her to marry him. Of course she accepted, separating from her lifelong friends and her New Orleans home to follow him to San Diego, where he had been reassigned by Convair. She also relinquished her career, though I suspect, in terms of passion, that she had surrendered this years earlier. In college, she had majored in chemistry, taught the subject to high school students, and gained admission to graduate school. But for reasons I never learned, she left her studies to work in the personnel department of a public utility.

When they married, my mother was thirty-five years old. My father had white hair. They had gotten a late start. But before Ike left office, they had begun a family, just like those of other, younger engineers. My birth announcement left no

My mother circa 1950.

doubt as to my father's profession: "Handicraft Limited reveals a few specifications on its new animated design model X2: Mary Grace," it began. "Wingspan: 16.8 inches. Length: 18 inches. Color: natural pink." It mentioned my gender coyly, as a "design feature."

"We regret that at the present time speed, range and altitude are confidential," it concluded. "However, we can assure you that she broke the sonic barrier on her first flight."

* * *

To understand JPL, you need to know not only what it was but also what it wasn't—just another aerospace contractor. In the 1950s and 1960s, the life of a rank-and-file engineer was itinerant. Aerospace companies had a binge-and-purge hiring policy. They would compete ferociously for big contracts, then, after winning them, take on droves of engineers, whom they would lay off when the project was done. The pattern at JPL—of hiring the best and having them stay—was atypical. In 1996, however, in response to the post-cold-war downturn in the aerospace industry, even JPL had to downsize, planning to cut twelve hundred of its six thousand jobs by the year 2000. I learned a lot about JPL from the engineers who discussed this in Donna Shirley's 1997 seminar. They weren't worried. They were insulted.

My father, by contrast, was used to instability. During my first three years in elementary school, Convair sent him for six-month stints to its Fort Worth, Texas, plant, as well as to a Lockheed plant in Georgia. I tracked down Ernie Kling, an engineer who had worked with him from 1950 to 1962, to discuss those years.

Kling reminded me of my dad. It wasn't just the whistles, chanting, and thuds emitted from his six-foot television, on which the Notre Dame football team was trouncing Navy. (Kling had graduated from Notre Dame in 1938.) Or the blue haze from the pipe he chuffed continuously. It was his slang. He referred to Convair headquarters as "the Rock." Designing ailerons and control surfaces was "working flap and slat." He cursed a patch of "wing-wong" sidewalk that had caused him to take a recent spill. *Wing-wong* means "awry," as in: "When you go supersonic, everything gets all wing-wong. You push the stick down, and you go up."

I asked Kling, who had been president of the engineers' union at Convair, a question that puzzled me. If engineers hated being laid off, why didn't they use collective bargaining to gain job security?

He rolled his eyes. "They think they're too smart," he said. They see themselves—and this harks back to the cold-war stereotype—as "'independent guys.' They'd join up when we were negotiating a contract, then quit."

My father, ever the "independent guy," had railed against unions.

Machinists, Kling continued, tended to support their union and to keep their jobs.

Kling had worked with my father on the PBY, as well as the CV-240 and CV-340 transport planes. My father was a liaison between the "floor," or shop, and the engineering division. When newly forged parts didn't fit together, he designed adjustments and conveyed them to the engineering division, where they were inked onto master drawings. The two men had also been involved with the biggest thing that Convair incubated during those years: the Atlas missile, or, more specifically, the airframe for the Atlas missile.

The Atlas was big not just in physical size but in its implications. Seventy-five feet long, ten feet in diameter, it was an intercontinental ballistic missile (ICBM), whose purpose was to transport a nuclear warhead to the Soviet Union. Its earliest version, containing three rocket engines beneath a thin metal skin, was static tested in 1956. Until 1954, the U.S. Air Force had favored traditional pilot-flown bombers as a means of delivering nuclear weapons. But in 1952, after "Mike," a thermonuclear device and the brainchild of physicist Edward Teller,

*My father and I celebrate Easter in our backyard in the early 1960s.
(The watch was an Easter present.)*

was successfully exploded at Eniwetok, a Pacific atoll, the air force changed its thinking. Because thermonuclear weapons were more destructive than atomic ones, they didn't need as precise a delivery system to eradicate their target. And they weighed less—enough less to be transported by a missile.

So while Kling's four children scampered through his avocado grove in the eastern tip of San Diego county, and my first-grade class read books by my neighbor, the author Dr. Seuss, near the northern tip of San Diego County, and countless other Convair families went their happy ways in other parts of the county, a vast delivery system for an object that could expunge an area the size of that county took shape, nurtured by our engineer dads.

The Atlas, however, was not intrinsically lethal. Blasting off as an Atlas-Centaur, with a second rocket, or stage, attached to it, it had an alternative identity. It was a launch vehicle, enabling many robotic probes, including Mariner Mars 69, to escape the Earth's gravity.

✳ ✳ ✳

In the early 1960s, "nuclear," too, was a positive word. Far from an incipient Chernobyl, San Onofre, the atomic power plant near San Diego, was a happy destination for fourth-grade field trips. Nor did "nuclear family" imply dysfunction. It was the ideal to which my parents aspired. But because both of them had come from fractured homes, they weren't confident that they could achieve it.

The daughter of German immigrants—my grandfather made beer at the Dixie Brewery in New Orleans—my mother lost her mother to cancer when she was nine. She was brought up, along with her six brothers, by her two older sisters, one of whom was so traumatized by raising teenage boys that she fled into a convent. (The other married but did not have children.) As a consequence, when it came to family, my mother didn't trust herself. She was always checking herself against other mothers, seeking validation. She didn't even trust her feelings when she went into labor. She believed her obstetrician, who told her by phone that what she felt could not possibly be contractions. He was soon proven wrong, however, when I arrived prematurely on the kitchen floor.

My parents studied guidebooks on family life. I still own one called *At Home* by Paul R. Hanna and Genevieve Anderson Hoyt. Published in 1956, it showcases a model family unit, whose nucleus is Dad. He asserts his primacy in a struggle over the television set, of all things. "Father is home, Susan," the model son tells the model daughter. "We have to see what Father wants to see now."

Of course in the 1950s, the whole country seemed conformist. No one I knew stood out or wanted to stand out. La Jolla, which today is a cosmopolitan university town, was then a backwater, its identity merged with that of San Diego, home to a naval base, whose tone was set by defense contractors and the military. As I child, I never questioned my parents' values. I'd like to say that I chafed at the uniformity, but the fact was, I loved it. I was a born hall monitor.

I was also in love with my father. Given my adult ambivalence, I assumed that I had been distant from him. But home movies belie that memory. Many are over-the-top Freudian pantomimes. One takes place in a powerboat bounding over Lake Tahoe. I

My mother and I pose for a snapshot in a neighbor's yard, 1961.

clamber onto my father's lap; he puts me down. I scramble up; he puts me down. When, finally, he lets me stay, my mother slides over and wraps her arm around us. Infuriated, I shove her out of the frame.

Another unfolds at a backyard barbecue. I climb into a tree house, lugging the decrepit space helmet that my father had rescued from Convair's trash. My mother follows me up and pretends that she is stuck in the child-size entrance. I play along, trying to push then pull her out. But when I realize that she has upstaged me—that the laughing onlookers are cheering for her—I turn away, put on the helmet, and slam the visor shut.

I learned about drama through such episodes, not from children's editions of Shakespeare. Literature, history, poetry, even current nonfiction were not priorities in our home or the homes of other engineers. What mattered was science. I was barely born when the Soviets launched Sputnik on October 4, 1957. But I felt its effect, the National Defense Education Act, which prodded kids to study science. My mother, who had dropped out of graduate school in chemistry, made one thing perfectly clear. I would not be dropping out of anything.

When it came to competition, my father was lower-key, perhaps because of his checkered academic past. (My mother had forced my father to finish his degree in night school.) Nevertheless, we spent many hours one year on my entry in my school's annual kite contest, a dazzling tubular object that somehow created lift. In another school, where everybody's dad was an investment banker or a corporate lawyer or a professor of Greek, my kite would undoubtedly have taken first prize. But this was San Diego, the land of aerospace, and the contest was a Skunk Works of cutting-edge concepts—multiple tubes lashed together, biplanes, triplanes, bat wings. My kite did not even earn honorable mention.

My father didn't seem to mind. Had I been his son, this might have been different. To a degree, both girls and boys were encouraged to dream cold-war dreams, to picture themselves as astronauts, soaring through space in defiance of the Russians. During the first years of the Mercury program, even the Barbie doll—that teaching tool for fifties femininity—was issued with a silver space suit. But in classroom filmstrips and movies, there was a puzzling gap between the hordes of girls depicted peering through microscopes and the total absence of grown-up women portrayed as scientists or engineers.

Recently, I screened several educational films from the 1950s and 1960s, including *Why Study Science?* (1955), which is typical of the genre. It features an idealized nuclear family—Mom, Dad, Betty, Jack—on a summer camping trip. When they look up at the stars, the conversation turns to science, which Jack plans to study so that he can travel to the moon. But his sister, he says, doesn't need science, since she "just wants to hook some guy."

If "hook some guy" isn't jarring enough for a classroom setting, it is followed by Betty's complacent response: "What's wrong with that?"

To which her mother rejoins, "Nothing. Except your idea that you'll never need science to keep house!" Or to prepare nutritious meals. Or to explain to your toddler how the telephone works.

"Suppose you don't find the right man right away," her father adds. Science can prepare you for a rewarding career as a nurse or a dental technician—jobs significantly less well remunerated and of dramatically lower status than, say, doctor or dentist. I began to understand Betty's complacency. "Hooking some guy" seemed a plausible alternative.

The film, whose soundtrack had been chirping crickets, suddenly explodes with patriotic music. "You and Betty and thousands of other young people . . . will have to learn all you can

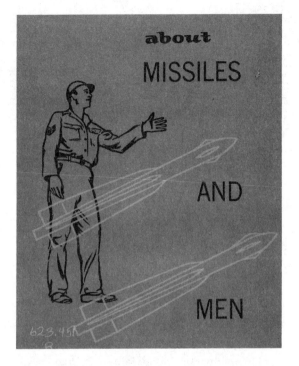

One of the cold-war-era books that made very clear which gender built missiles.

about today's technology," a narrator continues, to be leaders in your community (we see a male politician), to "heal the sick" (we see male doctors), to "build buildings and highways" (we see male engineers). The narrator concludes, "In the study of science is found the most useful knowledge of man."

Textbooks and primers from the period are no less dismissive of women. I got my first glimpse of JPL in a children's picture book from 1959 called *About Missiles and Men*. The book's cover says a lot about sex and rocketry, though at age eight I was too unsophisticated to decode its message. It shows two missiles darting across the page—one placed so that it appears to rise from a ground technician's groin.

You don't have to know semiotics, however, to figure out that women were not welcome in the field, except as menial laborers. "Many men have worked many years to make today's missiles," the text begins. "Maybe one of them is your father or your uncle. Maybe one of these men lives next door to you. Maybe you will be one of these men some day." Women appear in only one picture—hunched over long tables in a barnlike sweatshop that, a photo credit suggests, was at JPL. A Simon Legree with a plastic pocket protector oversees them. They ignore him, focused on the tiny parts in their tiny hands—or, in any event, their smaller-than-men's hands—parts that will be assembled into an electronic guidance system for a missile.

I could cite dozens of books and films with similar content. Viewing them was demoralizing, but, on the bright side, they gave a boost to my memory of my father. He had come by his misogyny honestly. He was a reflection of his time and his world.

Oddly, such propaganda hadn't put me off the idea that I could become a scientist—or a politician or a doctor or a writer. There was one thing, however, that I never wanted to become: a woman.

✳   ✳   ✳

If schoolbooks had made being a woman seem undesirable, events at home made it appear worse. When I was in third grade, my mother received a diagnosis of breast cancer and had to have a mastectomy. I was not sure exactly what these things were, but I knew that they were bad, and linked to being female. My mother and father discussed them behind the locked door of his study. She sobbed. I was afraid.

When she came back from the hospital, I felt as if I had two mothers, or maybe three. Around my father, she was almost maniacally cheery, and after several martinis, he, too, would brighten up. With me, she was subdued but not down, concerned that I not use her as an excuse to neglect school. But when I helped in the kitchen, I saw something that disturbed me. Not—as I might have noticed now—the relentless unwholesomeness of our steak-and-chops diet or the toxic maraschino cherries filled with red-dye-number-two. I saw frailty. She lifted full milk cartons with her left arm because she no longer had strength in her right.

When she thought she was alone, however, she was not bubbly. The surgery, I discovered, had sundered not just breast but underarm tissue; excruciating exercises were required to regain normal movement. I know this because I spied on her. The rupture to our perfect life ended my career as a hall monitor. I became a sneaky kid. What was the point of being good or adhering to rules if such behavior didn't stop bad things from happening?

As I questioned the beliefs I had been taught, my mother became more devout. Each morning she went to six o'clock Mass, often taking me. "Ask God to protect Daddy's job," she told me, but God had other plans. To engineering families, layoffs were like nuclear accidents; we knew they could happen, but we hoped and prayed and secretly believed that they wouldn't. When the ax fell at Convair, we were not prepared. At fifty-seven, my father was not the ideal job candidate. Mother exhausted herself cheerleading. Eventually, he secured "job-shop," or contract, work in Los Angeles, three hours north of La Jolla. For two years we saw him only on weekends.

From my perspective, this was not entirely bad. My father was a classic midcentury dad, permissive good cop to my mother's disciplinarian bad cop. She banned weeknight television; he was transfixed by Saturday-morning cartoons. She forbade me to build model airplanes because the glue was toxic; he arrived one Friday with a squadron of balsa-and-tissue kits. (We spent the next twenty-four weekends building Supermarine Spitfires, Grumman F6F Hellcats, Messerschmidt 109s, Mitsubishi Zeros.) Then there were chores. Helping my mother prepare, say, liver and onions, a slimy organ meat and an eye-stinging vegetable, was loathsome. Helping him whip up frozen margaritas was fun. At age eight I could mix a mai tai, a gimlet, a Singapore sling.

What I loved most, however, was when he talked about what he did. In the spring of 1965, he began work at Northrop Corporation on flight controls for the HL-10 and M2-F2, designs that had originated at NASA's Langley Research Center. They were wingless "lifting bodies," or wedges that could fly. The HL-10 had a flat stomach, the M2-F2 appeared pregnant, and both looked as if they would plummet like a brick when dropped from a larger plane, which is how they were intended to fly. Yet such wingless craft, he told me, were the future of spaceflight. You could launch them on a rocket, and they would glide back to Earth. They could shuttle astronauts to a space station. They would live in history.

As it happened, the M2-F2 did live in history, but not for those reasons. In May 1967, during a test flight at Edwards Air

Force Base, it went down in a spectacular crash—maiming but not killing its pilot, Bruce Peterson, the model for Steve Austin in the seventies TV show *The Six Million Dollar Man*. Footage of the actual smashup was broadcast each week during the show's opening credits.

Fortunately, the M2-F2 stayed aloft long enough for Northrop to offer my father a permanent job. We could finally move to live with him, a prospect that thrilled me until I saw our new home — Long Beach, California, a dreary community with a dreary oceanfront flecked with dreary oil rigs camouflaged as high-rise hotels.

My father didn't spend much time there, which may explain why he liked it. He was commuting to Pasadena for Mariner Mars 69. He worked long days, sometimes at JPL, often ferried there from Northrop by helicopter. To a sheltered ten-year-old, this seemed glamorous, worthy of James Bond.

✳   ✳   ✳

Before July 14, 1965, when JPL's Mariner 4 spacecraft provided the first close look at the planet, Mars enjoyed a homey, cozy reputation. "That Mars is inhabited by beings of some sort or other we may consider as certain as it is uncertain what those beings may be," observed American astronomer Percival Lowell in his influential 1906 book, *Mars and Its Canals*. Lowell's conviction stemmed from what astronomers saw: spotted dark areas that could be interpreted as vegetation and crisscrossed regions that appeared to contain canals. They also saw polar ice caps that expanded and receded with the Martian seasons. Even at midcentury, serious people expressed hope that a Mars trip could involve meeting Martians. In 1960, *This Week*, a general-interest national magazine, published a three-part novella by Wernher von Braun that detailed a human Mars mission. In it, twelve male astronauts powwow with intelligent beings living beneath the Martian surface.

Mariner 4, however, put a damper on such wishful projection. The patch of Mars that it photographed looked more like the moon than Venice, Italy. It was cratered, rocky, and bleak.

Scientists did not take well to this news; nor did the general public. People wanted another look. But there are constraints on travel to Mars. The periods when Mars and the Earth are best aligned for a trip occur roughly every two years. Nor are U.S. missions merely influenced by orbital mechanics; the machinery of Congress plays a part. In the mid-1960s, after successful missions to Venus and Mars, JPL began planning a "grand tour," a voyage not to inner-planets Mercury, Venus, or Mars, but to the outer planets, Jupiter, Saturn, Uranus, Neptune. In 1965, however, Congress yanked funds for this ambitious mission, leaving NASA scrambling to take advantage of the Earth-Mars launch opportunity in 1969.

Charles Kohlhase, a Georgia Tech physics major who was hired by JPL in 1959, had been a planner for an ambitious Mars mission that entailed two large spacecraft, each with an orbiter and a lander, to be launched on an immense Saturn V. In 1967, however, this project, which was to have been called Voyager, was scrapped, and Kohlhase was appointed the mission designer on Mariner Mars 69. (The name Voyager was later applied to a mission to the outer planets that was launched in 1977.) Although my father didn't know Kohlhase, he strikes me as the engineer my father would have liked to have been—technically brilliant, quietly iconoclastic, and knowledgeable about nontechnical things, such as photography and digital art.*

Kohlhase opened my eyes to what a mission architect does — or, more specifically, how a navigator gets a spaceship to a planet. A Mars probe requires enough energy to escape first from the Earth's surface, then from orbit around the Earth, and finally from the Earth's orbit around the sun. The navigator's goal is to place the spacecraft into a "transfer orbit," which will take it from the Earth's orbit to the orbit of Mars. Then, when Mars comes around, presto, the probe will be close enough to snap pictures of it or gather other data. (A ship intended to orbit Mars must fire rockets to slow its speed so Martian gravity can snare it.)

This is not easy. The Red Planet makes a complete trip around the sun every 687 Earth days. Earth completes its journey in roughly half that time. The trajectory analyst must get the spacecraft to the point where Mars will be when the spacecraft has traveled far enough away from Earth to encounter the planet. One former JPL navigator describes this process as "threading a needle from ten miles away."

The first thing a mission architect does is map out a "porkchop plot." This typically involves diagramming possible launch dates within a one-month launch period on a horizontal axis, and arrival dates on a vertical axis. The resulting set of curves earned its name because it looks like a line drawing of a pork

*Since his retirement in 1998, Kohlhase has gained recognition as a visual artist. He also serves on several space mission advisory groups, including the Mars Program System Engineering Team, a panel of experts that reviews decisions for JPL Mars missions.

chop. A mission architect can change the variables on this plot (or, in practice, overlays of multiple plots) to chart the optimal mission. Among the variables are the weight of the spacecraft, the energy capability of the launch vehicle, and the mission's scientific objectives, which could alter the weight of the spacecraft if it must carry heavy instruments.

Using this plot, mission designers and project scientists hash out where the spacecraft will go. The plot shows the limits of the craft's capacity: If it launches at a certain time with a certain amount of energy, it can only arrive at certain times and at certain locations. Because scientists know how Mars will be tilted and lit at these arrival times, they can direct the mission's navigation team to target the regions that interest them.

In 1966, JPL engineers determined that the spacecraft for Mariner 69 would be similar to the one that flew on earlier Mariner missions, an octagonal "bus" with four solar panels. But it needed to carry more elaborate and, consequently, heavier scientific equipment: a more powerful TV camera, an infrared photometer (to sense the surface heat of Mars), and an infrared spectrometer (to discern the spectral distribution of that heat energy).

This is where Northrop, the contractor, came in. Its mission was to shave weight off the body of the spacecraft, so it could convey the new instruments and still be light enough for an Atlas-Centaur to propel it to Mars.

✳   ✳   ✳

To discover what exactly my father had done on Mariner 69, I needed to talk to someone who had worked with him. I found a group photo of the Northrop team, but I didn't recognize anyone, and neither did my source—my father's former girlfriend, whom he began dating a decade after I moved to the East Coast. He met her while she was working in the Northrop art department.

She suggested that I ask about the picture at the monthly potluck and bingo game held by the company's retirees. She also suggested that I place a hibiscus in my hair. That month's potluck had a Hawaiian theme.

At the gathering, which took place in a community center near the Northrop plant, I showed my photo to dozens of white-haired men in flowered shirts and plastic leis. One couldn't hear me over the recorded strains of Don Ho. Another complained that I was interfering with the bingo calls. Eventually, though, I met Constantine ("Costy") Cafaro, who remembered my father. He identified several other men, and we planned to get together. Unfortunately, however, when I tried to follow up, he had unexpectedly died.

The mortality rate for men in the photo was high. I made dozens of false starts, trying to track down team members through the phone book, the Internet, and the Northrop-Grumman newsletter. I finally connected with Des Arthurs, who had been in his twenties when he worked on the project. Recently retired as Northrop's vice president for engineering, he lives in an elegant house in Palos Verdes and drives a Porsche, not your typical function-over-form engineer's car. (Kohlhase still proudly drives his 1986 Honda.) Arthurs was unique in other ways, too. He had grown up in Ireland and done his graduate work at Cranfield College of Aerospace, now Cranfield University, a British institution founded after World War II, whose students learned cutting-edge technology from Nazi "booty":

The Northrop team gathers in front of a model of Mariner Mars 69. (My father is in the back row wearing glasses; Des Arthurs is in the front row on the far left.)

aircraft, wind tunnels, rocket engines, test stands, even entire labs that had been hauled to England, part by part. Arthurs had risen to head Northrop's B-2 stealth bomber team. His bookshelves were filled with B-2 models and engineering awards. They also held a paperback of Virginia Woolf's *Mrs. Dalloway*, which he said he was reading.

Arthurs is a structural engineer, and his job on Mariner 69 was to make its parts light and strong. Strength mattered particularly during launch, when they were subject to violent shaking and the forces of acceleration. He showed me photos of tests at JPL that involved shaking a solar panel at different frequencies.

This wasn't so much to check the panel itself, but to check the accuracy of computer programs devised to model the panel's vibrations.

As "cognizant hardware engineer for mechanical devices," my father was essentially in charge of gadgets—things like actuators, joints, and the dampers that secured the solar panels during launch. He designed them and supervised their fabrication. When I suggested that this seemed slight, Arthurs reproved me.

In space, he said, nothing is slight. If a tiny part fails, the mission can fail.

On the drive back from the interview, I realized how right he had been. For example, although the Galileo spacecraft successfully orbited Jupiter and photographed its moons from 1996 to 2003, it did so with a handicap. Its high-gain antenna, which was folded like an umbrella during launch, never unfurled. A speck of lubricant, engineers believe, rubbed off of the antenna's joints on a trip from JPL to Cape Canaveral. The lack of a high-gain antenna crippled the spacecraft's ability to return science data at a high rate. During flybys of the moons, engineers had to record data on the spacecraft's tape recorder, then dribble it back through the ship's much smaller, low-gain antenna.

Because of a teeny glitch in a teeny joint, data that should have gushed through a floodgate trickled instead through a straw.

* * *

As I entered seventh grade, my mother was becoming an upbeat statistic. She had had no recurrence of cancer. She was vice president of the Charles Evans Hughes Junior High School

PTA, coleader of my Girl Scout troop, and a tireless gatekeeper against weeknight television. She forced me to make friends, afraid my love of novels would transform me into that pariah of pariahs, a "bookworm." Her right arm had become strong enough to pitch a softball. Which she did. At me. Appalled that I was never chosen for school teams because I invariably struck out, she spent a week in our backyard teaching me to whack the irritating ball.

My father, too, was upbeat, flushed with hope instead of martinis. But when he brought me to a JPL open house, I grasped the stepchild status of contractors there. Acquainted with many JPL men, he could talk inside talk with insiders. Yet despite this familiarity, he lacked the badge that would have branded him as one of the elect. He was as much a guest as I was.

Kohlhase, by contrast, was at the red-hot center of mission planning. "Most people on those early missions felt excited and privileged," he said. "None of us slept late. We were up at 6:30, eating breakfast quickly, solving technical problems we knew how to solve. We didn't take coffee breaks. We didn't watch the clock." Sometimes at lunch, to relieve the pressure, they would play "Kriegspiel," a version of chess where opponents sit back-to-back, deducing one another's moves without seeing the board. The image filled me with tenderness. My father loved mental challenges. I could picture him playing that game.

To the kids at school, the space race, too, was a game, but on January 27, 1967, we discovered just how high the stakes actually were. During a simulated countdown for the first Apollo flight, three astronauts were killed in a sudden fire. A spark ignited the pure-oxygen atmosphere in the command module. Gus Grissom, Ed White, Roger Chaffee—incinerated on the launchpad during a routine test. The tragedy spooked my class.

It also spooked my mother. She couldn't get over the speed of the accident or the effect it must have had on the astronauts' families. "Imagine those men," she said. "One minute they're on top of the world, on top of a rocket. The next minute they're gone." Didn't NASA know pure oxygen was flammable? Everyone knew that.

My father was philosophical. "Hard lessons," he said, "are the ones you never forget."

"Those families didn't need that lesson," she countered.

A few days later, on the eve of her fiftieth birthday, she herself underwent a routine physical exam. The doctors found colon cancer. It had progressed swiftly and without detection to an advanced state that required immediate surgery. Not as fast, of course, as a fire through pure oxygen. But very fast.

✳   ✳   ✳

"Did you see much of your family when you were working on Mariner 69?" I asked Kohlhase.

We were sitting on the floor of his house, where the light was best, looking at pork-chop overlays for the Mariner mission. His Shetland sheepdog, Ben, was trying to nuzzle the papers out of his lap and climb into it.

"With the long hours and all, did you see much of your wife?" I pressed. Then I became embarrassed. During the Mariner mission, he disclosed, his wife had left him—for someone more available.

After the separation, he saw his daughters every Wednesday and Sunday. "I never let them down," he said.

Without my pushing him, he continued, "I can't believe I weathered some of the things I weathered. Maybe you either implode as a human being or you survive it all. And it's close, which way you go."

He added, "JPL saved me."

*　*　*

Northrop delivered the two Mars-bound spacecraft to JPL on schedule. They were then shipped to Cape Canaveral for launches on February 25, 1969 (Mariner 6) and on March 27, 1969 (Mariner 7).

Although my father was no longer involved with them, he followed their progress, getting updates from Cafaro, who had been assigned to JPL to advise the flight team during the mission. My father's new project was a differential maneuvering simulator for NASA, Langley. A piece of ground-based hardware, it didn't have the same urgency as a spacecraft. But he kept long hours, as if it had.

I wish I could say that while he was gone I stood by my mother. But I was no better than he. I became editor of the junior-high yearbook because its staff met after school. Anything to avoid going home.

Worse, years of absorbing *Why Study Science?* and *About Missiles and Men* had brainwashed me. At puberty, I had begun to transform myself into an alien, coquettish creature. I made good grades quietly, on tests, but stopped sticking up my hand every time my teachers asked a question. I became that bundle of insecurity, cliquishness, and cosmetics obsession known as an eighth-grade girl. For older teenagers, 1969 was a time of defiance, of nonconformity, of taking a stand against the Vietnam War. But all I wanted was to be like everyone else.

Our family, however, was not like everyone else. My mother was very, very sick. By 1969, she had had a second operation and two brutal rounds of chemotherapy. The sisters who had brought her up came to our house to care for her. My aunts were not young and stylish like the moms of my friends. Aunt Anna was sixty-two, widowed, and a resident of Kilgore, Texas. Aunt Frieda was sixty-eight and a nun. Neither could drive.

During her visits, Aunt Anna imposed order on our house, which I welcomed, even as I fought against it. Aunt Frieda, by contrast, seemed to require more maintenance than she contributed. To lapse into the vernacular of the time, she creeped me out. I remember stumbling upon her in our kitchen, where she had drawn the blinds, shed her wimpole, and poured herself a tall martini.

I am not proud of things I did during those years. Few creatures are more craven, superficial, or socially ruthless than an eighth-grade girl. On rainy days, when I couldn't walk to school, our next-door neighbor took me. She drove a truck with a camper on the back, a tacky thing, in my eighth-grade view, in which I didn't want to be seen. I asked her to drive my mother's station wagon. When she wouldn't, I made her drop me off three blocks from the school.

I was sure that if the other kids knew how weird and wrenching and pain-filled my home life was, they would shun me. I tried to maintain a facade of normality at school, just as my father, for the same desperate, wrongheaded reasons, tried to maintain one at work. My relationship with him grew increasingly rocky. We each saw in the other what we despised in ourselves.

As spring became summer, the Mariners reached Mars. The eyes of the world, however, were on the moon, where, on July 20, American astronauts Neil Armstrong and Buzz Aldrin walked. Long Beach is an aerospace town, a bedroom community for McDonnell Douglas. The applause there was deafening.

We clapped and cheered in my honors summer science class, to which I had been invited because of my prior grades in science. I had enrolled in it, however, for a typical teenage reason. The boys in the course greatly outnumbered the girls. That July, my short-term objective was to persuade a geeky boy from a different school, whom I had met at a district-wide math contest, to kiss me. This potential occurrence seemed as implausible and miraculous as a trip to the moon. I had no long-term objectives. Because of my mother's illness, I could not imagine the future, much less plan for it.

On July 30, while the Apollo astronauts were still in postlunar quarantine, two things happened. My mother underwent a third operation — a hasty, emergency procedure in response to an infection. And Mariner 7, the second of the two Mars ships, malfunctioned. Both these events upset my father, but he would only discuss Mariner 7. He transformed his fear for her into fear for the spacecraft.

In fairness, the spacecraft's crisis was dramatic. He learned details from friends at JPL. Yet because he was not a member of the flight team, there was nothing he could do. He was as powerless to help it as he was to help her.

Kohlhase was in a different position. At the time of the problem, Mariner 6 had successfully swept across Mars's equatorial region and sent back seventy-five images. Mariner 7 was supposed to swing by the southern polar ice cap. But "in the last few days before the pass," Kohlhase recalled, "Mariner 7 had an onboard anomaly." Its signal disappeared. Navigators feared it had been struck by a meteoroid.

As mysteriously as the signal had vanished, however, it suddenly reappeared. The spacecraft's battery had sprung a leak, engineers determined. This resulted in a minor explosion that had pushed the spacecraft off its course and, worse, had confused the "orbit estimate," a prediction of where the spacecraft would be. Although engineers had pinpointed the problem, the ship was not yet out of trouble. The navigators had to figure out its location, then reprogram its science instruments, so that when it flew by Mars on its new course, the instruments would be pointing toward the planet.

Kohlhase was at the center of this fix. After the navigators had located the spacecraft, they fed its coordinates into a computer program that translated the changes in the spacecraft's orbit into changes in the pointing angles for its TV camera and other instruments.

"When the person who ran the program showed me the numbers, I didn't feel they were right," Kohlhase told me. "I didn't feel they matched the new orbit estimate. So I changed them in my head." Then he did something, he said, that "would never happen today." He sent his version of the commands—not the version from the computer—up to the spacecraft.

It was a hard decision. "Although the changes were small, I

JPL engineers Jim Campbell (left) and Charles Kohlhase work on a navigation problem for Mariner Mars 69.

aged a year in an hour," he recalled. "I remember sitting in front of a computer monitor, waiting to see the first pictures of Mars — knowing that if I never saw any, or even if they veered too far from the desired scientific swath, my career was over."

But the pictures arrived. One hundred twenty-six of them. Thirty-three taken during the spacecraft's closest approach. And they showed what the scientists had wanted to see: the planet's south polar cap.

There was hope in the world, there had to be. The summer had been full of evidence. Two men had landed on the moon.

The boy had kissed me. My father's spacecraft had made it to Mars. Miracles happened. Technology was unflappable. I permitted myself a guarded euphoria. Science could lick a medical problem.

By September, however, my euphoria had faded, as had the apparent euphoria of the Mariner science team. As feats of engineering, Mariners 6 and 7 were indeed triumphant. But the data they sent back were grim. The pictures looked as dead and deadly as those returned by Mariner 4. At a press conference in August, immediately after the second flyby, George Pimentel, a scientist at the University of California–Berkeley and the principal investigator on the infrared spectrometer, announced that his instrument had detected methane and ammonia, two gasses linked to the possibility of life. By September, Pimentel had to admit that he had made a mistake.

To ensure money for subsequent Mars missions, however, these depressing facts had to be presented with a cheerful spin. The September press conference on Mariner was a study in euphemism. Robert Leighton, a Caltech scientist and head of the team that interpreted the photographs, finessed the cratered bleakness. Mars, he said, does not resemble the moon or even a rock-strewn desert on Earth. Rather, Mars looks "like Mars."

There was, however, no way to put a bright spin on my mother's cancer. That fall we learned it had metastasized. In March, she entered Long Beach Memorial Hospital. Her doctors predicted she would live no more than a month. She survived six months, until the following September.

Recently, I found a snapshot of our family from March. We sit on a tired brown sofa. My mother is skeletal. She raises a glass of

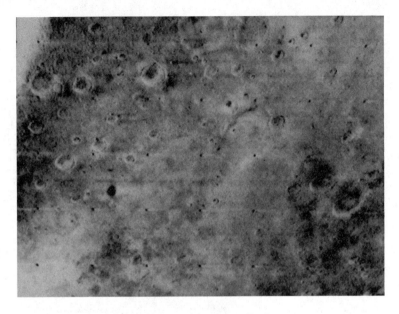

Mariner 7 took this image of Mars, near its south pole.

wine with her thin hand, the back of which is bruised from an intravenous drip. She smiles weakly. I am smaller than I remembered. Eyes shut, I grasp her tightly around her waist, like a sort of human limpet, as if by hanging on I could hold her longer in this world.

My father, however, looks most shell-shocked. There are pencil smudges on the front of his pale blue stay-pressed shirt. He sits apart from us, gazing at nothing, his eyes as black and vacant as outer space.

✳ ✳ ✳

My father soldiered on for another twenty-four years. Northrop forced him to retire in 1971 at age sixty-five, but he returned to do contract work on the FA/18. He remained employed until his late seventies. Work nourished him; it provided a goal, drinking buddies, the borrowed invincibility of bold men.

I escaped to college, at Yale, and, later, to New York City, where I had a job on a newspaper as a political cartoonist. Childhood prepared me for an aspect of this craft. You can mine nearly any grim event for humor.

My father and I stayed in touch, but at a remove, both geographic and emotional. Prolonged contact revived memories that we preferred to forget. We were not close again until 1993, when I spent time on the West Coast to do research for my book on the Barbie doll. Three NASA spacecraft had gone to Mars since Mariner 69. In 1971, Mariner 9 orbited the planet, and in 1976, two Viking landers touched down. (Mariner 8 had failed at launch.) Perhaps appropriately, during my first extended visit in twenty years, my father was agitated over a missing Mars probe. In August 1993, the Mars Observer, a JPL orbiter, had vanished only days before arriving at the planet. Brandishing a newspaper account of the disappearance, he was armchair-quarterbacking the failure investigation. Did anyone remember what happened to Mariner 7? he wanted to know. Had the solar panels been properly secured at launch?

He had retired in Fullerton, near my cousin, as well as near the Richard M. Nixon Memorial Library. We made a pilgrimage to see the tape recorder on which Rosemary Woods had done her erasing. We barbecued hamburgers. Before he died, also of cancer, I felt we had made peace.

Near the end of his life, he took morphine for pain. Where

Scotch had made him mute, morphine made him garrulous. It unlocked stifled emotions. He talked about the shame he had felt as a child, shame that made him demand burial next to his father, shame that he had never dared to confess. Shame, needless to say, for which a tough, silent, midcentury engineer would never have sought therapy.

He talked about my grandfather's accident. While in boarding school outside Baltimore, my father had been summoned to Boston but not told why. (His nuclear family lived near Chicago; his extended family near Boston.) He changed trains in New York City, where, as a thirteen-year-old boy, he figured he had time to take in a magic show. The show, however, ran late. He missed his train. Which caused him to miss his father's funeral, a mistake so disgraceful to him that he had never revealed it to anyone, not even his siblings.

This event took place seventy-five years earlier. But, he said, he had thought about it every day of his subsequent life.

✳   ✳   ✳

I thought of my father in October 2002 when NASA released a report from its inspector general's office that detailed fraud. One contract worker pleaded guilty to accepting twenty-seven thousand dollars of collect telephone calls from prison inmates and charging them to the Hubble Space Telescope program. Others had falsified test results, misrepresented repairs, and fabricated shoddy parts. Three NASA employees were arrested for stealing moon rocks.

To me, however, such malfeasance seems so foreign as to be,

well, Martian. My father cared deeply about his dampers and actuators. He viewed his job as a sacred trust.

Although their active life is over, Mariners 6 and 7 are still traveling around the sun, somewhere near the orbit of Mars. More than any terrestrial marker, they strike me as a fitting memorial to him, as well as to the other engineers who did their work with care. Unless they collide with a meteoroid, they will circle forever.

The Rockets' Red Glare, Part I

or

THE UNLIKELY BEGINNINGS OF THE JET PROPULSION LABORATORY

When I think of the nonengineering books in my father's study in the 1960s, three volumes come to mind: *None Dare Call It Treason* by John A. Stormer, *Masters of Deceit: The Story of Communism in America* by J. Edgar Hoover, and *A Choice Not an Echo* by Phyllis Schlafly. If there is any doubt about my parents' political conservatism, those books should banish it.

Stormer, who in 1962 "left a successful career as editor and general manager of a leading electrical magazine" to proselytize for the right, was the most global in his paranoia; he unearthed alleged sedition everywhere from UNESCO to the public schools. Hoover, director of the FBI, was the most manipulative; he told his readers scary stories, then recruited them to spy on the subversives next door. And Schlafly, best known for her seventies campaign against the Equal Rights Amendment, was the most complex. She did not simply pare the world into capitalist good and socialist evil. She located evil within her own Republican Party* and castigated the East Coast es-

---

*In 1960, Schlafly was president of the Illinois Federation of Republican Women.

tablishment "kingmakers" for not giving conservatives enough say in choosing presidential candidates. She did, however, manage at least one anti-Communist outburst—chiding President Lyndon B. Johnson for presenting the Enrico Fermi Award, "which carries with it a tax-free purse of $50,000," to physicist J. Robert Oppenheimer, who had "at least one Communist mistress."

I can't say what Schlafly, or, by extension, my parents, found most distasteful about Oppenheimer—that he had a mistress or that she was a Communist. But I suspect it had to do with communism. I revisited these books to understand the era's black-and-white thinking, and how it contributed to the harassment of JPL cofounder Frank J. Malina during the McCarthy period. Anti-Communist passion—and fear that Western technical superiority was at risk—also fueled the country's willingness to embrace Nazi scientists like Wernher von Braun, even as it repudiated native sons such as Malina. Although Malina may not today be a household name across the United States, his legend endures at JPL, where, in 1997, I first encountered him in effigy—part of a strange art installation that turns up in the lab's central courtyard during its annual public open house.

The display, which borders on folk art, features a group of age-ravaged department-store mannequins with a stack of sandbags, a glass canteen, and what looks like a bicycle pump with a thermos on top, to which hoses are attached. One figure wears a pith helmet with a spike at its crown. Another sports a faded vest, chipped eye, and hair that, due to excessive exposure, resembles a fright wig. To outsiders, the mangy group with the weird contraption is a strange sight indeed. But to those who know JPL history, it is a secular crèche—a dramatization of the lab's first rocket test—as resonant an image in science history as Ben Franklin tugging on his kite.

The display does not do justice to the men in the 1936 photo on which it is based. While it captures some of their raffishness, it doesn't convey their intensity or their extreme youth. Sprawled in the photo's right foreground is rocketry pioneer John Parsons, age twenty-two, the offspring of a wealthy Pasadena divorcée. Too distracted to finish college, he was a self-taught chemist whose investigations into rocket propulsion were guided by "Magick"—the sex rituals of self-proclaimed "Anti-Christ" Aleister Crowley—rather than mathematics. Behind Parsons is his childhood buddy Edward Forman, a mechanic who could cobble together almost any device out of junkyard finds. Apollo Milton Olin Smith, a Caltech graduate student known to his friends as "Amo," wears the pith helmet; its "spike" is actually a miniature fan that he designed to cool his head. Smith is bracketed on the left by Rudolph Schott, a student assistant of no lingering importance in rocket history, and on the right by Malina, who is of tremendous importance—the focused, visionary Caltech Ph.D. candidate in thermodynamics who served as a director of JPL.

The men surround their baby: a primitive rocket motor secured on a stand for a static test. One hose fed the motor gaseous oxygen; the other fed it fuel—in this case, methyl alcohol. The container atop the sandbags held water, pumped as a coolant into a jacket surrounding the combustion chamber.

On Halloween 1936, three miles from Pasadena's Rose Bowl, in a deserted stretch of canyon known as Devil's Gate Dam, the men first fired their device. This event did not mark the "founding" of JPL; it was, however, the founders' first em-

JPL rocketry pioneers sprawl near a rocket motor they will test in the Arroyo Seco, 1936. They are (from left) Rudolph Schott, Apollo M. O. Smith, Frank Malina, Edward Forman, and John Parsons.

pirical foray into rocket propulsion. The diabolical overtones of date and place may not be coincidental. As a disciple of Crowley, Parsons was one of Southern California's more notorious occultists—or, some say, Satanists. He served as leader of the Agape Lodge of the Church of Thelema, a chapter of Crowley's Ordo Templi Orientis (OTO). *Thelema* is a Greek word meaning "will," and the group's motto was "Do what thou wilt shall be the whole of the Law," pretty much the antithesis of the Golden Rule. Implementing one's will involved fornication rituals—a hodgepodge of tantric yoga and Christian blasphemy—that frequently took place in Parsons' home on South Orange Grove Avenue, which Pasadena residents referred to as Millionaire's Row.

If Parsons added a rakish bravado to the rocket research, Malina kept it on course. No less quirky than Parsons, he, however, had discipline. Where Parsons preferred to bumble along, concocting propellants by trial and error, Malina was a theoretician. A rocket motor that works on paper, he believed, should translate into one that works in life. He proposed a doctoral thesis on rocket propulsion and recruited Theodore von Kármán, Caltech's legendary Hungarian aerodynamicist, as his faculty adviser. This was no mean feat. In 1936, rockets were considered crackpot stuff. When Malina asked Clark Millikan, a Caltech professor and son of Caltech president and Nobel Prize–winner Robert A. Millikan, to serve as his faculty adviser, Millikan refused, suggesting dismissively that Malina seek work in industry.

With the outbreak of World War II in Europe, however, the military grew interested in solid and liquid propellant research. This was not, initially, to build a guided missile on the lines of the Nazi V-2, but to create a device that would help heavily loaded planes take off on short runways, such as those on aircraft carriers. These rockets were called JATOs, an acronym for "Jet-Assisted Takeoff," and were ultimately manufactured by Aerojet Engineering Corporation, a potentially profitable company that Malina also coestablished in 1942. The men used *jet* instead of *rocket* to refer to their device because rockets were still too far-fetched to be taken seriously.*

Malina's greatest accomplishment, however, came at the end of the war: He led the JPL team that designed and built America's first government-funded, liquid-propellant high-altitude rocket, the WAC Corporal. It was a "sounding" rocket—a high-altitude probe carrying instruments to report on the condition of the upper atmosphere—and a prototype for guided missiles to come.

By 1945, however, with Japan smoldering from two Allied atomic bombs, Malina recoiled from designing weapons. He broke out in "cold sweats," he told an interviewer, when he thought of the slaughter that a guided missile, delivering a nuclear warhead, could cause. Soon he had fled the war business, signing on in Paris, France, with an organization founded in 1947 to promote peace: UNESCO. But in 1953, he threw colleagues another curve, leaving UNESCO to become a studio artist. First painting, then sculpting, he engineered a system for creating a unique form of kinetic sculpture, which he exhibited throughout Europe. He continued to work for peace, focusing on what British novelist C. P. Snow had termed "the two cul-

---

*In 1920, after U.S. rocketry pioneer Robert Goddard had published an article in a Smithsonian Institution publication that dealt with a hypothetical moon flight, he was painfully (and erroneously) mocked by the *New York Times*. Believing rockets would not work in a vacuum, it said Goddard "lacked the knowledge ladled out daily in high schools." Decades of ridicule followed. Smarting from such slights, shortly before his death in 1945, Goddard wrote that the subject of rocketry must "be avoided in dignified scientific and engineering circles."

tures," which, by midcentury, were increasingly at odds—literary intellectuals and scientists. This crystallized in *Leonardo*, an international journal he founded in 1967, which deals with the interface between technology and art.

With accomplishments like these, you'd expect to find him enshrined in history, alongside Goddard, the quirky collector of rocketry patents who did not work well with others, or von Braun, the oily ex-Nazi who very much did. But a dark thread runs through Malina's résumé, as dark, for a time, in the national consciousness as Satanism itself—the allegation of being a Communist.

"My dad never talked about what he went through," Roger Malina, Frank Malina's son, said. An astronomer and physicist, the younger Malina holds the directorship of the Laboratoire d'Astrophysique, a French space institute in Marseille, and edits *Leonardo*. He strongly resembles his father in the 1936 rocket photo. They share an unexpected mix: piercing, razor-sharp intelligence that radiates from gentle, rounded features, which look as if they had been removed from the mold too soon.

"In 1954, my father lost his passport," Roger continued. "My mother was English, and we used to go on vacation and visit her family in Lancashire. But he never went with us. At the time I thought he just didn't like to travel."

It wasn't as if Malina had been singled out for harassment. Scientists with a social conscience were a pet target of Senator Joseph McCarthy, whose rampage against alleged enemies of the state took off in 1950, when, during a speech before a women's group in Wheeling, West Virginia, he brandished a list of suspected Communists in the administration of Harry Truman. McCarthy, however, was far from a lone voice of anticommunism. In 1954, accused by the Atomic Energy Commission of being a Communist sympathizer, J. Robert Oppenheimer, director of the Manhattan Project, lost his security clearance. The career of his brother, Frank Oppenheimer, who had been active in left-wing politics as a Ph.D. candidate at Caltech, was also derailed. Stripped of his teaching position at the University of Minnesota, Frank Oppenheimer left physics to become a rancher.

"The McCarthy committee targeted seven or eight people in UNESCO, including my dad," Roger recalled. "He quit UNESCO rather than being fired. My mother explains that for some period during that time, there was a French secret service car at the end of our street, watching everybody coming in and leaving our house. Many of our family friends—friends in the English Embassy, for instance—were told not to associate with my parents anymore. And some of them didn't—because their jobs were at stake. My mother said it was very strange. My parents had had a very active social life, and suddenly people refused to come to their dinner parties."

In *Enterprise*, published just before Malina's death in 1981, Jerry Grey, a former professor of aerospace science at Princeton University, wrote that because of Malina's Communist activities, he had been "forced to leave the work he loved and start over elsewhere." This, Roger said, made his father furious: He was not then, nor had he ever been, a Communist.

But Malina's first wife, Liljan Malina Wunderman, whom he divorced when he moved to Paris, remembers otherwise. "You have to understand how it was at the time," she explained. "Being a 'Communist' wasn't about overthrowing the government; it was about making sure people had things that they take for granted today: enough food, the right to unionize, health care, public education."

In his authorized history of the lab, *JPL and the American Space Program*, Clayton R. Koppes contends Malina was persecuted for participating in a "prewar political discussion group." Members read Marx and Lenin; they didn't pledge fealty to the Soviet Union. Yet this blurs a crucial detail: Sidney Weinbaum, the Caltech chemistry professor at whose home the group frequently met, was the head of a "cell"—Professional Unit 122—of the Pasadena Communist Party. In 1950, charged with lying about his membership in the party, Weinbaum was convicted on three counts of perjury and sentenced to four years in jail.

Far from Parsons's slapstick Satanism, Malina's story holds real intrigue. What were his actual allegiances? More significantly, how might those allegiances—or the cold-war-era perception of them—have corrupted the historical record? Linked to a major academic institution, JPL may have cherished its oddballs, but the U.S. Army didn't. And during the 1950s, the army had a major contract with the lab.

Worse, as part of Project Paperclip, a program instituted at the end of World War II, the United States welcomed Nazi scientists, some with problematic war records, who could be of use to the West, particularly in its budding space effort. To justify this, Malina believed, the United States had to discredit its homegrown scientific talent, especially those individuals who had flirted with Marxism. "The American government had an interest in showing that they had backed the right people," Roger recalled. So they had to deemphasize the American contribution—which is to say, the JPL contribution—to rocketry.

(Although a casual link is hard to prove, the deemphasis on JPL's accomplishments undeniably coincided with the hyperemphasis on those of the Germans.) "My dad," he added, "was very upset about" this.

Malina's fall is perhaps best appreciated in contrast to the blazing rise of the political chameleon Wernher von Braun. The two men were born in the same year, 1912. Yet where von Braun was bizarrely unscathed by his membership in both the Nazi Party and the brutal SS force, Malina was damned without a trial. Where von Braun mugged ceaselessly on American television—a 1955 show on space that he narrated for Walt Disney was so influential that President Eisenhower reportedly requested a copy to inspire his generals—Malina was shut out from the star-maker machinery. Where von Braun, observing the outspoken Protestantism of NASA's leaders, became a born-again Christian, Malina made no public professions of convenient beliefs.

Until 1945, Malina's passion served him well: His team almost single-handedly put American rocketry on the map. But in the history of the space program, engineering has always taken a backseat to politics. The space race was an invention of the opinion-shaping devices of the 1950s and 1960s: the shows we watched and the news we read. It was not about science. It was about outperforming the Russians.

Formerly classified FBI files, released through the Freedom of Information/Privacy Act, make it possible to see what was going on behind the scenes, when von Braun and his ilk were ascending in the foreground. Although the names of some informers remain blacked out, one can in large part discern who had it in for whom and how they exacted punishment.

* * *

Frank Malina was born October 2—a birthday he shared with
Parsons, though the budding sorcerer entered the world two
years later, in 1914. The astrology crowd makes much of this coincidence, yet for the roots of Malina's iconoclasm, one need
look no farther than his father. Born in Moravia, Czechoslovakia, in 1881, Frank Joseph Malina Sr. was a rebel: a Mason, an
agnostic, and a ferocious advocate of Czech self-government.
This did not endear him to authorities of the Austro-Hungarian
Empire, which had ruled Czechoslovakia since 1620. In 1848,
Czech nationalists had suffered a major setback: Their rebel-
lion against the Hapsburg monarchy was brutally crushed. Dis-
illusioned, thousands of Czechs poured into the United States.
Many settled in Texas, lured by rich farmland; the first organized Czech colony was formed in Galveston in 1852. By the
turn of the century, fifteen thousand Czechs had arrived, and
although they did not have the same impact on the Lone Star
State as its southern neighbor, Mexico, they left their mark. It
took the form of Bohemian and Moravian churches, Czech-
language newspapers, even the largely Czech polka festival in
Ennis, Texas.

In 1897, Malina's father joined that tide, immigrating to Brenham, Texas. But in 1918, when Czechoslovakia broke free of the
Hapsburgs and formed a republic, the elder Malina wanted to be
part of it. In 1920, he bought a hotel in a rural area of Czechoslovakia and repatriated with his family. Young Frank did not immediately take to his new life, particularly the part that involved
tending livestock.

Doing lessons in Czech, however, was no obstacle—a consequence of his facile intelligence. According to Roger Malina, the "gymnasium teacher in Czechoslovakia told his parents, 'This kid is really bright. You've got to go back to the States to find an education for him,'" which, five years after their departure, they did. In 1930, Frank graduated from Brenham High School (where his father, a musician, had secured a post as band director) and enrolled at Texas Agriculture & Mechanical University to study engineering.

This was not an easy choice. Frank had many talents. A precocious draftsman, he filled his boyhood room with his own drawings of flying machines. He had inherited his parents' musical ability, playing piano as a child and cornet in the Texas A&M band. Nor, as a boy, did he look exclusively to science for answers. "I took religion very seriously when I was a teenager," he told an interviewer. At age fifteen, he was confirmed in the Lutheran church, where his mother, a native-born U.S. citizen of Czech origin, played the organ every Sunday. "But my father was an agnostic, and this caused a crisis," he said. In college, however, after he read Charles Darwins *The Descent of Man*, his father's viewpoint won out.

Regardless of Malina's ultimate disposition toward Christianity, he obeyed the Fourth Commandment—the one about honoring father and mother. Nearly each week, from his departure for college to his final days (his mother survived him), he dispatched letters to his parents. They were filled with news, fears, confessions of romance, the suspense of experimentation, the exhilaration of proof. In those and other letters, he emerges as witty and self-deprecating, which may have been a studied pose. "He instructed everyone to save his letters for history," Wunderman remembers.

Like my father, Malina came of age in the Great Depression, the longest and deepest economic slump in U.S. history. But he didn't allow himself to be defeated by adverse circumstances. He recalls being so broke that he couldn't afford to travel to graduate school at Caltech. He managed to get there, however, by securing a loan from a Texas A&M professor. At Caltech, when he wasn't hitting the books, he scrambled for money — translating German papers, applying for fellowships, even illustrating an engineering textbook that von Kármán was writing. "Karman asked me why I am the only one in the department who can draw," he told his parents. "If I didn't need the cash you can bet that I wouldn't be drafting." In 1936, however, his fortunes improved. In recognition of his ability, Caltech awarded him a fellowship for wind-tunnel research that would cover tuition and provide three hundred dollars in additional expenses. The wind tunnel was part of the Guggenheim Aeronautical Laboratory at the California Institute of Technology (GALCIT), of which von Kármán was director, and which had been endowed by Daniel Guggenheim, a mining industrialist who funded aeronautical departments at various universities.

About that time, a speculative course on rocket propulsion that Malina took with Caltech professor William Bollay also began to pay off. This was not the school's only foray into daring, hypothetical territory. "We had seminars on all sorts of theoretical things," said Homer Joe Stewart, a Caltech instructor who joined Malina's team in 1938. Yet Bollay's subject caught the attention of a *Pasadena Post* reporter, whose story, in turn, attracted Parsons and Forman. In 1935, when they contacted Bollay, he put them in touch with Malina.

Frank Malina at Caltech, circa 1940.

By Feburary 1936, the odd trio—a Ph.D. student who defined himself as "suspicious of mystics," a mystic, and the mystic's slavish sidekick—were canvassing Southern California, searching for cheap parts with which to make a rocket motor. Although JPL commemorates it, the famous Halloween rocket test was, in fact, a bust. "The oxygen hose ignited and swung around on the ground 40 feet from us," Malina reported. To avoid being injured, "we tore out across the country wondering if our check valves would work."

The valves did indeed work, and by January 1937, after some revisions, so did the motor, proving that their concept could create thrust. It ran for forty-four seconds, long enough to convince von Kármán that the boys had gotten somewhere. He arranged for lab space on campus, in the basement of the Guggenheim building. Testing rockets, however, did not lend itself to shared quarters. The team's first major fiasco involved a rocket motor lashed to a fifty-foot pendulum that was suspended from the ceiling of the building's third floor. When the motor was fired, Malina figured, it would swing the pendulum, and the amount of swing would measure the rocket's thrust. Unfortunately, the motor, which used nitrogen oxide as an oxidizer, severely misfired—coating nearly every surface on three floors with a layer of rust.

Later, a motor affixed to a test stand blew up. It hurled a piece of jagged metal into the wall at the exact spot where Malina's head would have been had he not gone to von Kármán's house to return a typewriter. This incident sealed their banishment to the Arroyo Seco, the area where JPL is today. It also earned them a nickname: the "Suicide Squad."

If Malina joined Parsons in Satanic sex rituals, he was circumspect about such participation. To his parents, he characterized Parsons and his wife, Helen, as "good intelligent friends," adding that Parsons "reads a lot and has similar viewpoints on social problems as I have." In his memoir, *The Wind and Beyond*, von Kármán was more forthcoming, describing Parsons as a

*John Parsons, Edward Forman, Frank Malina and Lieutenant Homer
Boushey pose in front of Ercoupe, whose wings have been fitted with JATOs,
March Field, 1941.*

"delightful screwball" who "loved to recite pagan poetry to the
sky while stamping his feet."

The converse styles of Malina and Parsons were as essential to their motor as fuel and oxidizer. Yet they often led to
conflict. Malina struggled to restrict Parsons and Forman

from expensive, unnecessary rocket firings. Parsons and Forman, he grumbled, "are like inventors, in the worst sense of the word." Forman particularly chafed under Malina's curbs, and remembered them well into the 1950s. Bearing a grudge for events that had taken place in the early 1940s, he used his McCarthy-era FBI interviews to allege that Malina had held up wartime defense projects by fussing excessively with the underlying math.

Malina seemed most comfortable among scientists and high-culture buffs. He viewed von Kármán as a "second father." "My dad was very middle European," Roger explained, "as was von Kármán." Both loved music and chess and sought knowledge in a range of disciplines outside their field. "There was no pecking order of the sciences above the arts like there often is in American culture."

Malina was also smitten with the glamour of the von Kármán household. Back in Texas, his family had stuffed down an immense, meat-laden midday meal, which they metabolized by sleeping through the afternoon heat. Not so von Kármán, for whom evening meals were lavish, gracious social events, presided over by his sister, Josephine, called "Pipö," and, until her death in 1941, his mother, Helen. Besides teaching French at the University of Southern California, Pipö, who had earned a Ph.D. in art history, was the ultimate hostess, Pasadena's answer to Perle Mesta. With no visible exertion, she routinely transformed their house at 1501 South Marengo Avenue, with its ancient Persian rugs, Japanese screens, and Chinese tchotchkes, into a pageant of Depression-era glitterati. Caltech faculty rubbed elbows with artists, musicians, actors, screenwriters, and novelists. You never knew who would be at dinner:

novelist Theodore Dreiser, Danish physicist Niels Bohr, Italian physicist Enrico Fermi, or one of the Hungarian actors close to Pipö—Bela Lugosi, star of *Dracula,* or Jolie Gabor, mother of Eva and Zsa Zsa.

* * *

Ridiculed though it may have been, the Suicide Squad gained members. By 1937, it included Martin Summerfield, a geeky, myopic Brooklyn native and physics student, who lived in the same boardinghouse as Malina, and Hsue-shen Tsien, a Chinese Ph.D. candidate from a patrician family, whose aptitude for theory matched Malina's. Although his principal interest was rocket propulsion, Tsien, like Malina, was a musician. At parties, he occasionally played his recorder when Malina performed at the piano.

Malina put on a brave facade for his jeering classmates, but his letters home were a sine curve of self-doubt. The "rocket paper," he grimly reported in February 1937, is "getting longer and longer and emptier and emptier." Two weeks later, calculations he had made with Amo Smith suggested that the rocket would only reach a height of thirty-five thousand feet, when one allowed for air resistance, instead of the five hundred thousand feet he sought. Calculations made in late March, however, were more optimistic. They belied earlier results, and he forged on. (The work became the foundation not only of Malina's dissertation, for which he was awarded a Ph.D. in 1940, but of a second paper, completed in 1938, that won the REP-Hirsch International Astronautical Prize, issued by the Societé Astronomique de France.)

The team also struggled financially. When Weld Arnold, a photographer, offered to raise cash if he could shoot their experiments, cash was exactly what he raised: a bundle of one- and five-dollar bills, wrapped in newspaper, totaling five hundred dollars. (Eventually, he ponied up one thousand dollars.) Excluded from traditional funders by their work's eccentricity, they tried to keep an open mind. But most would-be backers, they discerned, were flakes. "Smith and I received a letter from a parachute jumper in N.Y. who wants to be shot up 1000 feet by rocket and then float down by parachute," Malina told his parents. "We wrote that we were sorry, but at present we couldn't help him."

By 1938, industry, in the form of Reuben Fleet, the president of Convair in San Diego, California, expressed interest in using rockets to lift heavily loaded planes. At Fleet's request, Malina drafted a paper intended to scare up seed money for JATOs. But Convair passed on the proposal. After a visit to GALCIT, however, General Henry H. "Hap" Arnold of the Army Air Corps requested that the Committee for Air Corps Research of the National Academy of Science give Caltech one thousand dollars for further study of JATOs. Clark Millikan, who had initially opposed Malina's rocket experiments, joined the team, and in 1939, their collective report so impressed the National Academy of Science that it awarded GALCIT a ten-thousand-dollar grant.

Work, Wunderman recalled, was always Malina's first priority; but he also had "his five-year plan," which included marriage. When in 1937, at a formal dance to which von Kármán had taken him, he met Wunderman, the former Liljan Darcourt, he didn't let her get away. She was a chic, French, seventeen-year-old art student, and he soon dazzled her, squiring her to concerts, plays, and soirees at von Kármán's. But in 1939, after they had exchanged vows in their living room at 1288 Cordova Avenue, he retreated into work. On a typical evening, after dinner, Summerfield or Tsien would arrive at Malina's, and the men would scribble calculations late into the night.

Although their union broke up in 1947, Wunderman remembers Frank as "a gentle, good man who really cared about people." Within the limitations of his era, he even had enlightened ideas about women. He encouraged Wunderman to finish her degree at Otis Art Institute. "Frank had no problem with a woman being an artist," Wunderman remembered. "As long as she made sure the coffee cups were washed when his friends came over.

"Looking back, I don't think Frank's and my marriage had a chance," Wunderman said, shaking her head.* "I was too young, and Frank was too busy." Yet according to FBI informants, both she and he found time in 1938 and 1939 for what the FBI considered Malina's first transgression: They attended meetings of Professional Unit 122 of the Pasadena Communist Party.

The JATO team, known as "GALCIT Project 1," worked long hours: 7:30 a.m. to 5:30 p.m., six days a week. Conditions were, at best, primitive. "When I first came to the laboratory," recalled Dorothy Lewis, the first secretary of GALCIT Project 1, I was "in this little shack" with "this little bitty switchboard." Also known as Malina's office, the shack "was so small you

---

*Wunderman did indeed become an artist who today shows her work as Jan Wunderman. I interviewed her in her Manhattan studio amid dozens of her abstract expressionist canvases.

could hardly move." Outside were dirt roads that became mud pools when it rained.

As war heated up in Europe, the engineers' focus intensified. "Kármán, with a Jewish background, was extremely sensitive to the Nazi development in Germany," Malina recalled. "And since I spent so much time with him, I also viewed the persecutions with horror." Breakthroughs came swiftly. In 1940, von Kármán and Malina proved it was possible to obtain stable burning in a long-duration solid propellant rocket engine and gave design criteria for stability. Within a year, Parsons had implemented their design, using compressed black powder (the stuff you find in Fourth of July fireworks) as a fuel. The JATO units were small—two feet by five inches in diameter—and were designed to produce twenty-eight pounds of thrust for twelve seconds, enough, they believed, to lift a light plane.

On August 12, 1941, at March Field, near Riverside, California, the men fitted a light monoplane called an Ercoupe with two of the three-unit JATO assemblages under each wing. By itself, the plane weighed 753 pounds; with a pilot and rockets, 1,023 pounds. All the key players were present: Malina, Parsons, Forman, von Kármán, and Summerfield. They were joined by Lieutenant Homer Boushey, a strong-chinned, dashing, former student of von Kármán's, who would pilot the Ercoupe. Although the Halloween rocket photo had been taken a mere five years earlier, Malina looked as if he had aged a decade. He and von Kármán appeared somber, resolute, shielded by Panama hats from the fierce desert sun. No longer was Malina a kid with a clever toy. He was a man, with the weight of a war on his shoulders.

The tests had begun with small steps. First the JATOs were fired on a stationary craft; then they were fired from a plane after

*Legendary Caltech aerodynamicist Theodore von Kármán
performs a last-minute calculation before testing the
JATOs, March Field, 1941. Malina (left) and Boushey
look on.*

the plane had achieved flight. But this was the moment of truth.
Boushey climbed into the cockpit and taxied down the runway.
As the plane accelerated, he hit the JATO ignition switch. Smoke
erupted from the rockets. The Ercoupe wobbled forward, then
rose sharply off the field—the first such jet-assisted takeoff in the

When the JATOs ignite the Ercoupe ascends from March Field.

United States. By the end of the day, after several successful take-offs, Boushey was so confident that he nailed another first: He flew the Ercoupe under rocket power alone.

The JATOs in the Ercoupe tests, however, were fired shortly after they were assembled. To the men's distress, they discovered, this was the only way the JATOs would work. If you stored them for more than a few days, they would blow up at ignition.

Parsons, as lead chemist, was pressed to find an untraditional solution, which he did: using asphalt as a fuel and potassium perchlorate powder as an oxidizer.* Some historians say the revelation struck while he watched a roofer lay down tar; occultists theorize that he channeled the idea. But wherever it came from, it solved the storage problem. GALCIT also forged ahead with liquid propellants. With storability in mind, the team settled on a combination of aniline, an oily, poisonous basic liquid, as a fuel and red fuming nitric acid as an oxidizer. "We now have something that really works," Malina, euphoric, blurted to his parents on March 20, 1942. "We will be able to give the fascists hell."

* * *

The fascists, meanwhile, were incubating JATOs of their own. Between 1941 and 1942, the German Rocketry Research Center at Peenemünde, directed by Wernher von Braun and located on the Baltic Sea, had designed an egg-shaped JATO intended to lift hulking German bombers. (The device was strapped under each wing, fired, then parachuted to a landing for reuse.) But the rivalry between the German Luftwaffe and the German army, to which the Peenemünde group reported, kept the JATOs grounded. Instead of implementing the Peenemünde design, the German Air Ministry encouraged Schmidding, the German manufacturer, to make "improvements." There was, however, one catch: The Schmidding "improvements" didn't work.

Malina and his team avoided conflict with a manufacturer by making the JATOs themselves, insofar as this was legally possi-

---

*The fuel in a rocket motor burns when it combines rapidly with an oxidizer. In a jet engine, by contrast, the fuel burns when it combines rapidly with the oxygen in the air.

ble. On March 19, 1942, for tax reasons, they chartered Aerojet Engineering Corporation in Wilmington, Delaware, with an office in Pasadena. Its president was von Kármán; its treasurer, Malina; and its secretary, Andrew Haley, von Kármán's lawyer. Parsons, Forman, and Summerfield served as vice presidents. As a nonprofit institution, GALCIT couldn't make money off its discoveries. But Aerojet could.

The fledgling company, however, was not yet out of the woods. It was one thing to lift a dinky Ercoupe off the tarmac, quite another to raise a twenty-thousand-pound bomber. On April 15, 1942, at Muroc Army Air Force Base (now Edwards Air Force Base), Malina's team put its new liquid-propellant, two-hundred-pound-thrust JATOs to the toughest test yet: lifting a hefty Douglas A20-A bomber. (These JATOs produced about seven times as much thrust as the ones tested on the Ercoupe.) Major Paul H. Dane, another former student of von Kármán's, was in the cockpit. The plane rolled forward, hesitated, then surged suddenly upward—as, in 1943, with a large navy contract for liquid-propellant JATOs, did the fortunes of Aerojet.

Prosperity stunned Malina. He also seemed to feel guilt. It took the form of edgy, self-conscious jokes: "We are getting to be more and more like capitalists," he observed after a golf date with Summerfield, "at least on the surface." To his parents, he remarked: "An accountant went through our books today and found nothing that could send me to Leavenworth."

Guilt and self-criticism, by contrast, did not seem to be in Parsons's emotional repertory. Yet oddly, for a guy whose motto was "Do what thou wilt shall be the whole of the Law," Parsons seems to have been habitually walked upon by others. By the mid-1940s, Wilfrid Smith, his predecessor at the helm of the local Agape Lodge, had run off with Parsons's wife, Helen. Undaunted, Parsons took up with his wife's sister, Sarah. But his heart belonged to another, more ethereal gal, "Babylon the Great, the Mother of Harlots and Abominations of the Earth," aka the Scarlet Woman in the Book of Revelation. He attempted to incarnate her through a regimen of ritualized sex, known to occultists as the "Babylon Working," or, more precisely, the "Babalon Working." (Aleister Crowley, Parsons's mentor, was a fiercely original speller.)

It is unclear whether the fornicating brought Babalon (or Babylon) to Earth. What can be documented, however, is that around this time, two problematic presences entered Parsons's life: L. Ron Hubbard, a science-fiction writer and the future founder of the "church" of Scientology, and Cameron, Parsons's soon-to-be wife, a vermilion-haired free-love advocate who occasionally generated what she considered to be works of art. Nor did Parsons's antics endear him to the rest of Millionaire's Row. It wasn't just the all-night chanting or the naked, narcotized women lurching around the garden; it was his goats. "He kept them in the backyard," Wunderman recalled. "You can imagine the smell."

The war, however, yanked even Parsons out of his self-absorption. Beginning in June, 1944, Germany showered Great Britain with its most malignant weapon yet: the V-1, or "buzz bomb," a pilotless, jet-propelled plane containing two thousand pounds of explosives. By September 1944, the Nazis pounded England with an even more potent weapon: the V-2, a long-range liquid-fuel rocket that carried about seventeen hundred pounds of explosives. Psychologically, the V-2s were the most disturbing to their victims. Because they traveled faster than the speed of sound, they arrived silently, in advance of an audible warning.

Although the Nazis managed to hurl missiles across the English Channel, their technical know-how did not exceed that of Malina's group. In 1943, Malina, von Kármán, and Tsien had submitted a proposal to the Army Ordnance Department titled "The Possibilities of Long-Range Rocket Projectiles." It sketched out a design for a U.S. guided-missile program, which the army agreed to fund. In the fall of 1944, Malina flew to England to examine a V-2 that had been downed over Sweden. But the hardware, he said, taught him nothing. "We were well along in our fundamental research by then," he told an official Caltech interviewer. The German technology had no "serious impact on the research program."

On July 1, 1944, GALCIT officially changed its name to the Jet Propulsion Laboratory. This reflected both a new purpose—guided-missile work—and a new structure. Von Kármán was no longer available to run the laboratory; he had been whisked off to Washington, D.C., to found and head the Science Advisory Group of the U.S. Air Force. In lieu of a single hand on its tiller, JPL was run by a governing board, whose acting chairman was Clark Millikan. Malina was named JPL's acting director, with Louis Dunn, another von Kármán protégé, as Malina's deputy.

As a first step toward a guided missile, the team planned to build and test a high-altitude "sounding" rocket, the WAC Corporal. Although Malina had been thinking about such a rocket since 1936, the WAC Corporal did not spring full-blown from his head. Three smaller versions were designed and tested, to see how various components would work. The first of these was the Private A, an eight-foot missile with four twelve-inch tail fins that flew suc-

A Private A missile takes off from Camp Irwin in California's Mojave desert.

cessfully at Camp Irwin in California's Mojave Desert. The second was the Private F, a Private A with "wings," two five-foot lifting surfaces where the tail fins had been. In tests at Fort Bliss, Texas, however, the Private F flopped: It corkscrewed out of control, demonstrating that the wings were a bad idea. The third version was wingless: the Baby WAC, a one-fifth-scale model of the WAC Corporal, which flew effectively at Goldstone Range, California, in July 1945.

Because of its larger size and wider reach, however, the WAC Corporal would be tested at a new site, the White Sands

Proving Ground in White Sands, New Mexico, a spot that Millikan and Stewart had chosen in the late spring of 1945. It was close to Mexico, a short drive from Ciudad Juárez. Indeed, after Milikan and Stewart had selected the site, Stewart says, they "went across the border into Juárez, saw a bullfight, and had a fine dinner."

✳ ✳ ✳

In Europe, by early 1945, Germany was clearly headed for defeat. Nobody's fools, von Braun and his team connived to save their hides. On May 2, 1945, five days before their country officially surrendered, they gave themselves up to the U.S. Army at Garmisch, a Bavarian ski resort. Had they capitulated at their research center at Peenemünde, or at Mittelbau Dora, the concentration camp where the V-2s were assembled by slave labor, they would have wound up in Soviet custody.

As part of his work for the U.S. Air Force, von Kármán was among the Allied representatives that liberated Mittlebau Dora, aka Nordhausen, in the Harz Mountains. He described the hollowed-out mountain as "a monstrous place" where the Nazis executed their prisoners by "controlled starvation." He explained: "Each man was given so many months to live. The calories in his diet were determined and then reduced at a set rate until he died of hunger. All this time the Nazis worked him in the factories. This was the most horrible thing I had yet heard of—a perversion of science beyond anyone's nightmarish imagination."

Nevertheless, the U.S. Army rushed to embrace the men responsible. Not only did it offer von Braun and his team jobs, it instituted Operation Overcast—later called Project Paperclip—to ease their passage into this country. Overcast began on July 6, 1945, when the Joint Chiefs authorized a secret project to "exploit . . . chosen, rare minds whose continuing productivity we wish to use," according to author Christopher Simpson in *Blowback: The American Recruitment of Nazis and Its Disastrous Effect on the Cold War*. And in 1947, when prosecution for war crimes at Mittelbau Dora began in Germany, the army saw to it that none of von Braun's men were sent back to stand trial.

In the 1960s, I was taught that von Braun, Arthur Rudolph (project manager of the U.S. Saturn V program), Walter Dornberger (instrumental in the development of U.S. ICBMs), and the other Germans admitted to the United States had been Nazis in name only and had not perpetrated war crimes. Rather, they were scientists who made compromises to realize their dreams of spaceflight. Recent scholarship, however, particularly *The Rocket and the Reich*, by Michael Neufeld, curator of World War II history at the National Air and Space Museum, Smithsonian Institution, strongly refutes this. Neufeld details the horrors suffered by inmates at Mittelbau Dora: the starvation diet, the routine beatings by members of the SS, the latrines made from oil drums cut in half and topped with boards, and the total absence of wash water. (Some inmates resorted to washing themselves with their own urine.) Not only was the stench terrible, but the threat of cholera was so great that the German civilian workforce had to be immunized against it. Inmates who did not succumb to cholera died of pneumonia, tuberculosis, dysentery, typhus, starvation, and physical exhaustion—so many that, between October 1943 and March 1944, "Dora's death toll was effectively 6,000 in six months." Any engineer or scientist involved in missile production, Neufeld points out, had to be aware of these conditions, if not actively complicit in them. In recognition of the deaths, Neufeld offers this assessment of the V-2 as a unique weapon: "More people died producing it than died from being hit by it."

What is more, Neufeld observes, defenders of von Braun, Rudolph, and Dornberger have claimed that these men confronted members of the SS with the need for better conditions at Mittelbau Dora. But such claims "must be regarded with the greatest skepticism, especially as there is not a single document to back them up. At most, Rudolph, Dornberger and von Braun argued that missile quality was not going to improve if the labor supply was not in better shape."

Neufeld did, however, unearth a document that showed von Braun handpicking slaves to build his missiles. In a letter von Braun wrote to Albin Sawatzki, production planner at Mittelbau Dora, von Braun described visiting the Buchenwald concentration camp "to seek out more qualified detainees," that is, detainees with technical educations, and arranging for them be transferred to Mittelbau Dora.

In 1985, the *New York Times* reported that "American intelligence officials concealed the Nazi records of hundreds of former enemy scientists to try to get them into the United States after World War II, contrary to a Presidential order and against the objections of the State Department, according to declassified government documents."

Some of these scientists, such as Rudolph, were eventually charged with war crimes. But others were investigated and never charged. Most chilling for me, with my memories of that discarded pressure helmet, was Hubertus Strughold, chief scientist

Wernher von Braun at JPL during the Mariner Mars 69 mission.

of the aerospace medical division at Brooks Air Force Base in San Antonio, Texas, who is known for his work on pressurized space capsules. During World War II, Strughold served as director of the Medical Research Institute for Aviation at the Aviation Ministry in Berlin. He was investigated because of a letter in a magazine that named him along with associates in connection with experiments on concentration camp inmates. The experiments were related to space science: They involved freezing some people to death and subjecting others to simulations of high altitude—low pressure and thin air—to test the human ability to withstand extreme conditions. Strughold claimed to have learned about the experiments after the war, but documents unearthed in the 1970s showed otherwise. He lectured on the temperatures of the sea at a 1942 conference where the results of the freezing experiments caused what the prosecution at the Nuremberg trials called a "sensation" among scientists.

Ralph Blumenthal, the *New York Times* reporter who during the 1970s and 1980s covered efforts to deport alleged Nazi war criminals living in the United States, was not surprised that the investigation had been abandoned. Strughold had connections to the Central Intelligence Agency, Blumenthal said, and "the government closed ranks, protecting the scientists on the space program. Because we were competing with the Russians, the value of the space program overrode any compunction we might have had in dealing with these people."

✳ ✳ ✳

In October 1945, after extensive debriefing in Europe, von Braun and his fellow Nazis were shipped to Fort Bliss in El Paso, Texas. Thus, when the JPL team began tests on the WAC Corporal at White Sands, the former fascists were a mere thirty-five miles away.

Accustomed to plush surroundings, the Nazis referred to their quarters at Fort Bliss as a "rat shack." But the U.S. scientists at White Sands bivouacked in conditions that were equally Spartan. "We are living with Army men in Army style," Malina told his parents in September 1945.

The primitive environment did not impair Malina's progress. On October 11, 1945, the WAC Corporal soared to a height of 235,000 feet—the first man-made object to escape the Earth's atmosphere. Significantly, this took place a good six months before the Germans achieved similar results at White Sands with their recycled V-2s.

✳   ✳   ✳

Because JPL deputy director Louis Dunn had been a protégé of von Kármán, one would have expected him to bond with Malina. But Malina kept him at arm's length. Dunn was born in South Africa in 1908. Although he became a naturalized U.S. citizen in 1943, he never renounced his white supremacist views. "I soon realized that Louis Dunn's attitude toward things that I thought were important, like racial equality, was quite different," Malina told a Caltech interviewer. But "we never argued about it." Rather, "there were territories we just left alone." Oddly, Dunn was not a swaggering Neanderthal but a quiet one, who seemed especially mute around women. He was "a brilliant man but hardly said a word," Lewis recalled. "I couldn't understand him when I took notes for him." He refused to look at her, preferring to mumble into the floor.

Still, no amount of tact could ease the tension with Clark Millikan, which began nearly a decade earlier when Millikan refused to serve as Malina's faculty adviser. "Clark Millikan hated Frank and everything he stood for," Wunderman said. This wasn't because Millikan thought Malina was a Communist; he merely believed Malina was a liberal. Rather, as Wunderman saw it, Milikan felt his "superiority of money and upbringing" had made him a better person than Malina, whom he dismissed as a rabble-rousing "commoner."

By 1946, Malina's conflict with Millikan had begun to eat away at him, even as another, greater conflict gnawed at his conscience. The atomic warheads that had flattened Hiroshima and Nagasaki in August 1945 were delivered by airplane. In the future, though, such weapons might well be dispatched by guided missile.

"I was getting caught up more and more in trips to Washington and meetings with the Army, Navy, and Air Force, planning the war," Malina told an official Caltech interviewer. "I made a report on the possibilities of rocket propulsion for the future to a committee of General Stilwell.* He, of course, thought it was all a lot of nonsense. But I found in these meetings, I was getting more and more disturbed, and I would break into cold sweats. I just hated the idea of, say, planning to use all this for bombarding people. . . . I felt the Second World War was unavoidable and that Nazis and fascism and these crazy ideas had to be defeated. But many of us hoped at the end of the Second World War that there was a chance, maybe through something like the United Nations, to put some kind of a control on the sovereign states to put a stop to war — at least between industrialized countries."

During the Second World War, Malina's patriotism was above reproach. "He is out to win this war and is solidly behind the war effort," an informant told the FBI in 1942. When World

---

*General Joseph "Vinegar Joe" Stilwell, commanded the U.S. forces in China, Burma, and India during World War II.

War II ended, however, Malina's environment changed. The great moral death match against fascism evolved into a show-down with communism. As a consequence of this change, Malina was perceived differently, even though he himself had in no way altered his behavior. By the middle 1940s, Communists, or "Masters of Deceit," as FBI director J. Edgar Hoover called them, were demonized in broad strokes. To a country that viewed international politics as a clash between teams—us and them, right and wrong, good and evil—Communists were them, wrong and evil. What is more, people who had expressed curiosity about communism in the 1930s were not allowed to re-consider. Regardless of the way their sympathies may have evolved, they were inexorably tainted.

Malina's first major brush with Hoover came in 1946, when he was away from Pasadena. Wunderman received an unexpected phone call from scientist friends at Caltech. Placing themselves at risk, they warned her that FBI agents had planned a surprise search of the home she shared with Malina. "Get out now," the friends told her. "They're on their way." Panicked, she ran to a bedroom where Malina had kept books with Marxist themes. Then, forcing herself to slow down, and at least try to be thorough, she scooped up the material and hid it in her car.

"I drove around for hours with the books," she recalled. "I couldn't just throw them in a canyon; they would be found, and some of them might have had names in them." When night fell, she lugged the books to an art-school friend's house in the Hollywood Hills. There, she, her friend, and her friend's mother built a bonfire and incinerated them.

Wunderman did not exaggerate. The first entry in Malina's FBI file is dated November 28, 1942. In it, an informant alleged that in 1941 Malina had "passed out Communist literature" to GALCIT workers. From 1942 until 1944, the FBI investigated him rigorously, but finding no evidence that he had remained active in the Communist Party, it backed off. Two years later, however, suspicious that he might have failed to disclose his Communist Party membership on a government security questionnaire he had filled out in 1941, it resumed intense surveillance.

In August 1946, Malina and Summerfield delivered a paper to the Congress of the International Union for Theoretical and Applied Mechanics in Paris. Their subject was "The Problem of Escape from Earth by Rocket." In it, they proved the feasibility of a new idea: a multistage payload-bearing rocket that could achieve escape velocity from the Earth. The trip also suggested an idea to Malina: escape from the FBI by relocation.

A year later, after applying for a leave of absence from JPL, Malina visited Albert Einstein at Princeton University, where Einstein was on the faculty. "Like almost all scientists," Malina wrote his parents, Einstein is "disturbed over the hysteria that is current in the U.S." It was another scientist, however, who pointed a way out. Julian Huxley, the charismatic English biologist and first director general of UNESCO, offered Malina a job in Paris.

Divorced from Wunderman, Malina threw himself into UNESCO with the same fervor he had brought to rocketry. As a Ph.D. student, to earn money, he had assisted von Kármán on a project on soil erosion for the Soil Conservation Service of the U.S. Department of Agriculture. They had studied, among other things, the aerodynamics of sand-dune formation. At UNESCO,

Malina returned to the desert, founding and heading the Arid Zone Research Program. As a metaphor, this assignment resembled one of those biblical retreats during which a transformation occurs—a sort of fasting in the wilderness before beginning one's ministry. And, in a way, the stint was. Malina had entered the UNESCO chrysallis as a divorced man and a scientist. He would emerge as a family man and an artist.

In 1948, his private life stabilized. He began dating Marjorie Duckworth, a colleague at UNESCO, whom he described to his parents as "English, very fine, and full of common sense." A year later, they married. "As the ceremony was in French," he joked, "we are not sure who is to obey whom." His son Roger arrived in 1950, though to Frank's alarm, he "looked like a little frog." By the next morning, however, the frog had become a prince, and even his mother appeared "chirpie," or, at least, willing to go through the process again. Two years later, she gave birth to a second son, Alan.

Challenged at work, happy at home, and freshly transplanted from a cramped apartment in the center of Paris to a modernist house on its outskirts, Malina seemed to have eluded McCarthy. But back in the States, the cold-war winds continued to rage, and by the summer of 1952, they had whipped through Paris. "UNESCO is being strongly attacked by the Catholics," Malina wrote his parents. Other critics included "the Daughters of the American Revolution—a fine combination."

The Vatican and the DAR did not have a problem with UNESCO's relief work. What nettled them was that the UN, unlike the U.S. government, refused to purge employees with suspected links to the embodiment of godless evil, communism. "In the *Washington Times Herald* for January 22, 1951, Fulton Lewis Jr. stated that there were ninety Americans employed by the UN whose records of Communist and Red front activities bar them from working for their own government," according to a memo in Malina's file. Malina, the memo continued, was one of them.

Other entries place a seditious spin on his every move. In September 1948, Malina attended and addressed a meeting of the World Scientific Workers in Prague—in part, as a pretext for a social visit with his Czech cousins. The FBI, however, dismissed the whole trip as participation in "a Communist front."

During Malina's visit to the United States in 1949, not only did agents shadow him; they speculated on his suspicious New York City mailing address: 405 East Sixty-second Street. Did this link Malina to the "Oxygen Equipment Manufacturing Organization" located there? Could this be a Red front? As it happened, however, the address was a typo. Malina picked up his mail at 405 East Forty-second Street—the UNESCO office at the United Nations.

Some pictures in his file are so distorted that they're funny. One, taken before his alleged Communist involvement, is brightly lit; between his hairless cheeks and his soft face, he looks like a choirboy. The second, taken after he was accused of being a Communist, is starkly lit; between his mustache, goatee, and dark, hooded eyes, he looks like a medieval portrait of Satan.

By 1952, the slapstick G-men had moved in for the kill. An FBI report dated January 4, 1952, citing a "reliable" source, asserted that in 1939 Frank Malina, using the Communist Party pseudonym Frank "Parma," had been issued Communist Book No. 1020—his alleged official party ID. Three days later, another report noted that the Personnel Security Questionaire (PSQ) that Malina had signed on March 26, 1945, had vanished from his Caltech file.

No one alleged that Malina's involvement with communism lasted beyond 1939 or, at the latest, early 1940. Wunderman suspects that Malina, like herself, was so disgusted by Soviet premier Joseph Stalin's nonaggression pact with Hitler in 1939 that he left the party, as she herself had done at the time. (FBI records indicate that Professional Unit 122 dissolved before the United States entered the war.) But whereas von Braun got to bury his Nazi past, Malina was allowed no second thoughts.

For almost six months, the FBI searched for the missing PSQ. In August 1952, after locating a duplicate copy through military records, FBI representatives marched into the office of Angus McEachen, the assistant U.S. attorney in Los Angeles, and asked him to indict Malina for fraud against the government. McEachen reviewed their "case" and determined that there wasn't one. He gave two reasons. First, he found "insufficient documentary evidence regarding Malina's membership" in the Communist Party. And second, "an omission to state membership" in the party on a PSQ was "not sufficient basis to conduct a prosecution for perjury."

Irritated but not ruffled, the FBI redoubled its efforts, pressing previous interviewees to incriminate. Presented with a huge batch of allegedly new material, most of which was the same old stuff phrased differently, McEachen caved. On October 10, 1952, he agreed to "authorize prosecution" and "proceed to secure an indictment."

The FBI's push to indict Malina had less to do with efficiency than desperation. The statute of limitations on Malina's alleged crime expired December 31, 1952. McEachan got a "true bill" of indictment from a federal grand jury on December 18, 1952. The indictment, which was kept secret, was posted on December 30, 1952 — one day under the wire. Nor would the indictment have happened, McEachan told an interviewer, without "considerable interest in this matter by the Bureau."

On February 10, 1953, Malina submitted his resignation to UNESCO. This surprised the FBI and, one suspects, frustrated those who had planned to use his case to discredit the agency. Malina next "tried to get political-refugee status in England," Roger said. But his luck had worn thin: "The English wouldn't take him."

The bureau declared Malina a "fugitive" and ordered his arrest if he set foot in the United States. Popular wisdom held that all Commies were spies, but the FBI could find no evidence that Malina had engaged in or planned to engage in espionage. Indeed, by 1953, he had not had access to sensitive technical material for six years. If he was prosecuted for fraud, the only charge that might remotely stick, this would, however, embarrass the UN. And embarrassing the UN, to those who wished to end its tolerance of former Communists, was a laudable goal. The FBI looked into extradition.

Because the case against Malina was "weak," Hoover himself told the bureau's L.A. office in a memo dated February 18, 1953, that the investigation "must be immediately intensified." Assign the case, he ordered, "to an experienced agent who has . . . the aggressiveness . . . to get out and 'dig.'"

*   *   *

McCarthyism took down Tsien and Summerfield, too. After the Chinese Communists seized power in 1949, the FBI accused Tsien of being a spy—a far cry from his earlier, ultrapatriotic status as a member of von Kármán's elite advisory panel to the U.S. Air Force. Although Tsien professed no desire to return to China, the Immigration and Naturalization Service held him under house arrest, then had him deported—a decision they may have regretted when he went on to found the Chinese ICBM program. (Not to mention in 2003, when the Shenzou 5 spacecraft transported the first Chinese astronaut into orbit, and Tsien, age ninety-two, was celebrated as the father of the Chinese space program.)

Summerfield, also alleged to have been a member of Professional Unit 122 of the Pasadena Communist Party, escaped to Princeton, where he became a professor of astronautics. But he lost his security clearance, which hampered his research.

Parsons retreated from Aerojet, sinking deeper into the morass of his private life. In 1945, his pal L. Ron Hubbard absconded with Parsons's girlfriend, Sarah Northrup. Besides raging against Hubbard and Northrup, Parsons blustered against McCarthyism. "Public employees have been subjected to ignominy and indignity of 'loyalty' oaths and 'loyalty' purges," he ranted in *Freedom Is a Two-edged Sword*, an essay collection published by an occult press. This "burlesque investigation," he added, "will cause pain and sorrow to many innocent persons."

Parsons may have known—about causing pain, that is. When interviewed by the FBI, he did not appear to behave with heroic reserve. In Malina's file, an informant fitting Parsons's description placed Malina at Communist Party meetings in Malina's house. On the other hand, the FBI considered this informant a

nutcase, and initially his testimony was not taken seriously. The
witness, wrote the FBI agent who conducted the interview, is "a
'character,' an eccentric," and of "questionable" reliability. Only
later, when the FBI was desperate to nail Malina, did they use
this evidence.

Parsons's opportunity to incriminate others was, however, limited. In 1952, at age thirty-seven, he was killed in an explosion
that may or may not have been an accident. Historians of the occult, wishing to inflate Parsons's importance, have attempted to
invent a conspiracy around him, pinning his death on antagonists
ranging from rival cult leaders to aviation magnate Howard
Hughes. But William Pickering, JPL's director from 1954 to
1976, who had worked with Parsons, suspects the accident was
due to "carelessness." Parsons often left volatile chemicals lying
around his kitchen, a danger Malina himself had remarked upon,
and which kept him from visiting Parsons's home.

After the end of the cold war, it is chilling to review
McCarthy-era FBI files. One finds friends and colleagues ratting
on each other, the sort of thing you associate with the Stasi in the
former East Germany, not with affable engineering types in
Southern California. But loyalty oaths and witch-hunting often
compromised "good" people. One way to experience the time—
and its almost implausible terror—is to read *A Season of Fear*, a
novel published in 1956 by blacklisted screenwriter Abraham
Polonsky. Few engineers wrote memoirs—especially not memoirs that dealt with persecution by the FBI. But Polonsky's fiction is a convincing evocation of an engineer's actual worldview.

The book plunges the reader into the world of Charley Hare,
a well-liked, seemingly honorable engineer. Hare is a not a bad
guy, but neither is he courageous, especially after he signs a loy-

alty oath and learns how witch-hunts work. In a world where allegations *are* convictions, outward appearances mean everything. "This was the great scientific secret about facts," Hare observes, "that they could look like any opinion you had of them as long as the opinion wasn't a scientific law." Soon, to protect his image, Hare is burning books and betraying friends. At the novel's climax, he conveniently allows an antagonist to drown while swimming.

Hare doesn't build rockets; he is a civil engineer. Yet Polonsky captured the thought processes of his engineer characters, which reflected the biases of their training. In school, engineers learn to solve technical problems, not to examine the moral and philosophical implications of those problems. When an engineer in Polonsky's novel encounters a colleague who won't sign a loyalty oath, the engineer brands the colleague "a do-gooder," which "no engineer should be." Then he adds, "We don't care who runs what or how, we have to do the job. But he"—the do-gooder— has a "nose and he takes his nose and sticks it."

When I read those lines, I thought of Frank Malina—and the way his social conscience must have riled many of his lockstep colleagues. Indeed, in Malina's real-life dossier, after being interrogated by the FBI, an engineer described as Malina's "friend" told his interlocutor that he "had given considerable thought to his association with Malina" and "did not intend to continue the friendship."

Malina's life, however, did not end with his indictment or with repudiation by his so-called friends. It took an extraordinary turn—a turn far stranger than fiction.

The Rockets' Red Glare, Part 2

or

PORTRAIT OF THE ARTIST

If Frank Malina's life were a movie, with a traditional three-act structure, this would be the end of act 2. Covertly indicted, singled out for harassment by J. Edgar Hoover, Malina was on a seeming fast track to ruin. Then fate dealt him an amazing card—not to mention a card that was richly ironic. At the height of his persecution for alleged membership in the Communist Party, an organization committed to the overthrow of capitalism, capitalism made him very rich.

In 1952, the General Tire and Rubber Corporation acquired Aerojet. Von Kármán, Parsons, Forman, Haley, and Summerfield sold their shares when this happened. Malina, however, kept his, and by 1953, they had trebled in value—reaching, FBI files indicated, about four hundred thousand dollars. And their value continued to rise.

The French authorities were not indifferent to wealth. "My father initially had some very unpleasant interviews with the French secret service," Roger said. "They wanted to move my parents to the country so they wouldn't have contact with anyone." When the French found out that Malina was well on his way to being a millionaire, however, their tune changed. They kept him under surveillance but permitted him to remain in Paris.

"My dad clearly wasn't going to be a burden on the French state," Roger explained. The FBI, one learns from Malina's file, wasn't sure if Malina's alleged offense was a ground for extradition under French law. The French, apparently, did not think it was.

Freed from having to earn a living, Malina began to study painting with Reggie Weston, an English artist with whom he had played chess.* He attended afternoon classes at the Academie Julien: sketching nudes, training his hand and his eye. While at UNESCO, he had kept tabs on the changing fashions of visual art, praising Picasso less for his innovation than for his business sense. "He certainly does turn out stuff in quantity," Malina wrote in 1948. Not "too excited about the substance" of Picasso's ideas, Malina was, however, oddly confident about his own, and his competitive spirit kicked in. "I want to look a bit more carefully into the art game," he wrote in 1953.

As Malina grew more proficient with traditional media, he became less interested "in making oil paintings of bowls of fruit," his widow, Marjorie, told me. He was as much his own man in art as he had been in science and government, and he approached art like an engineer. What he sought to achieve was no less than a sense of movement, painting with light itself.

---

*Many engineers, I suspect, have harbored secret ambitions to be artists—Sunday painters like my father for one, or, for another, former JPL engineer Charles Kohlhase, who works with digital imagery.

Art was defined narrowly in 1953. Even works that were supposed to have dealt with movement, for example, Marcel Duchamp's painting Nude Descending a Staircase, were a sequence of static images. Although Duchamp built some sculptures with moving parts around 1913, it was not until 1931, when Alexander Calder began assembling his trademark wire mobiles, that the art establishment paid attention to moving objects.

Thinking like the engineer that he was, Malina tried to find ways to expand the limitations of paint. By imposing a wire or string over the surface of a painting, he discerned, he could add literal depth (rather than merely its illusion) to the picture plane. When he showed these experiments in October 1953, at the Galerie Henri Tronche, they did not upend the Paris art world. Some of his reviews, Malina remarked, were "real stinkers." But the exhibition threw a curve to the FBI, which suspected, with no supporting evidence, that his artwork might be a cover for espionage. Agents filed a pamphlet about the show in his dossier and noted with seeming incredulity that the Museum of the City of Paris had bought one of his pieces.

A month later, in Los Angeles, the U.S. attorney who had secured the indictment against Malina had yet more second thoughts. After weighing new evidence involving the way the statute of limitations had been applied to Malina's case, he petitioned a judge to dismiss the indictment. On April 30, 1954, the judge did just that. Yet the dismissal did not free Malina from FBI scrutiny. Rather, with increasing puzzlement, agents continued to monitor what was going on in Malina's artwork.

In January 1955, while taking down the family Christmas

tree, Malina had an artistic epiphany. Since the Tronche exhibition, he had experimented with wire mesh, using multiple layers to create texture and depth. By incorporating blinking Christmas-tree lights in his wire thicket, however, he was able to add a new dimension: movement. The resulting object, which featured pulsing lights under tiers of wire, he called *Jazz*. The English art critic Reg Gadney, championing Malina's work in *The London Magazine*, was more generous. He called the object a "natural conclusion of Mondrian's attempts to simulate movement in his two masterpieces, *New York City* (1942) and *Victory Boogie Woogie* (1944)."

Soon Malina began "painting" with electric motors. These drove wheels that were divided into colored sections. Images were formed through the overlap of colors and shapes and were in constant flux. The apex of Malina's technical innovations may be what he termed his *Lumidyne* system. It involved light projected onto and through Plexiglas screens to form fluid, soft-edged, ever-changing patterns. Malina's Lumidyne pieces often evoked astronomical phenomena. Not dependent on external illumination—electricity drove their lights and motors—the radiant compositions suggest swirls of gas and dust around distant stars.

Although Malina's art had admirers, such as Gadney, it also had detractors, and he tried not to let his bad reviews get to him. "History shows that a strongly negative reception by professional critics is best for the artist," he wryly observed, "Especially after he is dead." Yet if the European art establishment was slow to embrace Malina's art, the FBI was slower.

On November 13, 1956, agents in Paris summoned Malina for an interview, grilling him about events that had allegedly taken

Frank Malina creating a kinetic painting, circa 1955.

place eighteen years earlier. Although Malina was "courteous and affable," the report states, he "declined to furnish any information," particularly when asked about other alleged Communists. "He was motivated," the report adds, "by personal scruples in refusing to identify such persons and requested that it be made a matter of record that his refusal should not be interpreted as evidence that he had engaged in Communist activities."

Malina's combination of courtesy and silence exasperated the FBI. Agents were also baffled by his social circle. Inspired by von Kármán's gatherings in Pasadena in the 1930s, Malina brought together all manner of artists, scientists, and intellectuals to eat, drink, play chess, and discuss issues in his Paris house.

"We had painters, cosmonauts, Dad's Czech cousins, you name it, visiting us," Roger recalled. "I can easily imagine some poor FBI guy looking at the people coming to our house, thinking: What is this place? One day, there's William Pickering, the head of JPL. The next, there's Leonid Sedov"—the Russian academic who, along with Sergei Korolov, was the architect of the Soviet space program. "Dad just went on as if there was no cold war." Indeed, the Malina guestbook is a sort of left-wing *Who's Who*. Autographs include *Ascent of Man* author Jacob Bronowski (a regular visitor), writer Mary McCarthy and her diplomat husband, James West, even folk singer Joan Baez and her former husband, draft resister David Harris.

Frank always was a "vivacious provocateur," said Sandy Koffler, a longtime friend of Malina's and the founding editor of the *Courier*, a UNESCO publication. But beneath his bluster— almost provocation for its own sake—there was decency, generosity, and compassion. "There were two words Frank hated," Marjorie Malina said. "*Randomness* and *cruelty*." As a scientist, Malina often railed against anthropomorphism, the projecting of human consciousness onto animals. Yet you would be hard-pressed to find someone more attentive than Malina to his fellow creatures. Not only did he grind up food for the wild birds in his yard; he engineered and built a wooden "runway" for them to land and take off. Nor could a dog owner have been more devoted to his dog than Malina to his cocker spaniel, Rocket. At dinner, when Malina ranted against anthropomorphism, Koffler joked, Rocket would "be under the table licking his hand."

* * *

Because he could not travel, Malina managed to avoid the early 1950s in the United States—a blessing perhaps, since he would not have enjoyed them. From 1946 to 1954, while Dunn served as director of JPL, the lab produced the Corporal and Sergeant missiles, capable of delivering a nuclear or conventional warhead up to eighty miles away. Nor was the nascent space program shaping up as the vision of international cooperation that Malina had dreamed it would be.

In the late 1940s, the average American had little knowledge of his or her country's achievements in rocketry. This is because there was no need to publicize them: Congress had not spent billions of dollars in peacetime for their funding. By the 1950s, however, the government had to rally support for vast expenditures in weapons research. This involved making the public aware of the perceived need for such work—an agenda with which many purveyors of popular culture were only too happy to comply. It wasn't that the government paid off the media in some sinister fashion to produce this impression; rather, those who controlled the content of the news—Henry Luce, for example, the owner and founder of *Time* magazine—were fierce opponents of communism. Critiques of McCarthyism rarely surfaced in movies, television, and similar mass cultural products. To be sure, many interpreted director Don Siegel's 1956 movie, *Invasion of the Body Snatchers*, as a parable against McCarthyism. Yet the film's prologue and epilogue render it ambiguous, less an indictment of Red-hunting than a broad commentary on paranoia.

The cartoon entrepreneur Walt Disney had an unambiguous anti-Communist agenda and a potent means with which to sell it. In 1955, his Sunday-night program, *Disneyland*, which ABC had inaugurated a year earlier, was the country's most popular show, with an estimated viewership of at least 42 million people. To understand the deemphasis on Malina in space program history, as well as the acceptance of the program's ex-Nazi leadership, one must visit the Disney workshop of the 1950s and observe the elves within, specifically those responsible for "Man in Space," a space infomercial that was broadcast on Disney's show, March 9, 1955, and "Man and the Moon," which aired December 28, 1955. The elves were Willy Ley, a science writer, Heinz Haber, a physiologist, and Wernher von Braun—all Germans. Ley and von Braun had, in fact, originally met as members of Berlin's rocketry and spaceflight society, the *Verein für Raumschiffahrt*, in the 1920s. (Unlike von Braun, Ley fled Germany when Hitler took power.) Working with and for Disney, these men saw to it that their efforts—German efforts, rather than those of the JPL pioneers— were enshrined in popular history as the first steps to the moon.

Disney didn't have to dig deep to find the Germans. They came to his attention—or rather, to the attention of Ward Kimball, the animator and producer whom Disney had delegated to prepare his programs on space exploration—through a March 22, 1952, cover story in *Collier's* magazine. Titled "Man Will Conquer Space Soon. Top Scientists Tell How in 15 Startling Pages," the special section featured pieces by Ley, Haber, and von Braun. It also included articles by two American scientists: Fred L. Whipple, a Harvard University astronomy professor, and Joseph Kaplan, a physicist at the University of California at Los Angeles, who were not invited to join the Disney brain trust. *Collier's* magazine claimed to have a circulation of 3 million, but von Braun's notoriety was not limited to its readers. Although his article had been heavily edited—his English was still not great—he was anointed spokesperson for the section. This paved the way for star-making appearances with such small-screen celebrities as Garry Moore, Dave Garroway, and C. Fred Muggs, a chimpanzee. What is more, the *Collier's* article was the first of a series, as well as the basis for a pair of popular science books, all of which kept von Braun in the public eye.

Hooking up with Mickey Mouse, however, may have been von Braun's most cunning move. The naturalized Nazi was a genius at self-promotion. In 1958, as part of a carefully crafted autobiography that he allowed to be serialized in the *American Weekly*, the country's most widely read Sunday newspaper supplement, he promulgated his own genesis myth, omitting such pesky details as what von Kármán had termed the "monstrous" conditions at the Mittelbau Dora concentration camp.

But even for von Braun, navigating the entertainment industry was tricky. *I Aim at the Stars*, Columbia Pictures' 1959 movie based on his life, backfired. People who had experienced the war in Europe did not buy Hollywood's sugar-coated telling. Londoners, who had a firsthand memory of Nazi V-2s, took to the streets to protest the film, as did droves of New Yorkers, including a large number of Holocaust survivors. In Antwerp, which had been the target of over one thousand V-2s, city authorities would not permit the movie to be shown.

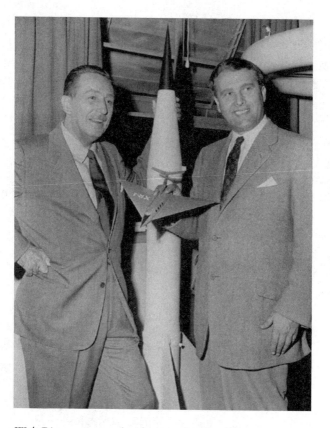

Walt Disney, creator of such pro-exploration TV programs as Man in Space *(1955), with his favorite technical adviser, Wernher von Braun.*

It wasn't as if Disney had always been a friend of rocket science; he only hopped on the bandwagon when space had become a cold-war playing field. Twenty years earlier, Disney could not have cared less about rockets or science—a fact that poor Malina, in the wrong place at the wrong time, had learned firsthand. In 1938, von Kármán, set to give a scientific talk to a general audience in London, got the idea that cartoons could help him convey tough technical concepts. He asked Malina to approach Disney, believing the animator would be willing to illustrate his talk—not for money but "for the sake of science." But Disney, who had never heard of the Hungarian, dismissed Malina. Not only would his animators not furnish drawings; he would prosecute any unlicensed use of his mouse.

In the middle 1950s, Disney's fiefdom was, of course, far from the global empire that it is today; but its influence within the United States may have been even more intense. Besides television, Disney controlled another powerful promotional tool: Disneyland, the theme park he opened in Anaheim, California, in 1955. Here, too, the German team imposed their version of history. Disney divided his park into "lands": Main Street, Frontierland, Fantasyland, Adventureland, and Tomorrowland. Main Street and Frontierland celebrated idealized fictions of America's past. Fantasyland was a commercial for his studio's animated fairy tales. Adventureland offered a taste of imperialism; on rides like the Jungle Boat, suburban white folk could gape at mechanical aboriginal people. It was in Tomorrowland, however, that Disney did his best to imagine the American future in the shape of the German past.

Tomorrowland's main draw was a ride called the Rocket-to-the-Moon, which used projected images and a vibrating theater to create the illusion that its occupants were traveling through space. A giant rocket emblazoned with the Trans World Airlines logo stood outside the theater. Instead of making this rocket in the image of the U.S.-produced WAC Corporal, Corporal, or Sergeant missiles—a simple tube shape with a pointed nose—however, Disney immortalized the pride of his German advisers. The rocket has curves and embellishments reminiscent of the V-2.

Some of Disney's Germanophilia may have been innocent. For example, he patterned his Fantasyland castle after Neuschwanstein, King Ludwig II's loopy, turret-ridden Bavarian fortress. To a degree, illustrators of 1950s science fiction favored the sweeping curves of the V-2 and had made them part of the era's iconic rocket. But when you consider that the V-2s had hammered London less than ten years earlier, the oversight—or choice—still seems bizarre.

On October 4, 1957, however, the embryonic space program became less reliant on its propagandists. A slackjawed American public gaped skyward in disbelief. The Soviets had launched a satellite, Sputnik I, before we could get ours off the ground. A month later, they shot off Sputnik II, which contained a dog and no provisions for the dog's return to Earth. Soon, Americans feared, they would send up a nuclear arsenal.

Von Braun was quick to smell fear and quicker still to exploit it. In 1958, he arranged to testify before the Elliott Committee, a congressional committee on education and labor. "The Communists' threat to the free world extends far beyond armies and politics," he told Congress, fanning paranoia with each inflammatory word. "It involves every aspect of our way of life: religion, economics, industry, science, technology, and education." The soul-denying state system was "turning out competent engineers and scientists in greater numbers than ours." Hence it was imperative that the United States "recruit more young people into scientific and technical careers." Congress could afford to spare no expense. American engineers, students of engineering, and, most important, rehabilitated Nazi engineers must be given all the money they needed to halt the Communist juggernaut.

* * *

Explorer 1, the first successful American salvo in the space war, was fired January 31, 1958. It was launched by a four-stage rocket of which the top three stages, as well as the instrument-bearing satellite, were not the handiwork of former Nazis. They had been designed and built at JPL.

After the launch of Explorer 1, space became so thoroughly a cold-war theater that it was impossible to imagine nations cooperating up there—unless, of course, you were a dogged internationalist like Malina. When, in November 1964, Malina advanced a plan for an international laboratory on the moon, the FBI viewed the project as subversive. Agents also suspected that the International Academy of Astronautics, which Malina had cofounded in 1960, had a seditious agenda. Like UNESCO, the group endorsed international partnership rather than U.S. supremacy.

Malina presented his plan for a lunar laboratory at the Third International Symposium on Bioastronautics and the Exploration of Space in San Antonio, Texas. (Having had his indictment dismissed in 1954, he could by this time return to the United States without fear of arrest.) Afterward, he swung by Brenham, Texas, to visit his mother. The FBI followed. Oddly, it was one of Malina's high school classmates—Tieman Dippel, the president of the Farmers' State Bank in Brenham, and, for twenty-three years, the sheriff of Washington County — who put the bureau's wackier suspicions to rest. Malina, Dippel assured, was "thoroughly loyal," "patriotic," and "trustworthy."

On November 12, 1981, while working in his studio outside Paris, Malina died of a sudden heart attack. Shortly thereafter, most of his papers were shipped to the Library of Congress. But Roger Malina remembered a strange object that his mother had recently unearthed in Paris: a sketch for a screenplay written by Malina and Parsons in 1937. The screenplay deals with the relationship between his father's rocket group and the Nazis.

Located behind an ancient, vine-covered wall at the end of a cul-de-sac in Boulogne Sur Seine, Malina's house is not exactly on the beaten track. Yet when one passes through the front gate, which has to be unlocked by hand, the appearance of a remote fortress disappears. The front yard is cozy, dominated by a build-it-yourself log cabin that was once the headquarters for *Leonardo*, the international journal of art and technology that Malina founded in 1967. The structure now holds his grandchildren's toys.

Inside the house, a vast picture window floods the living room with light, so intense that Malina designed a heavy curtain to protect his art collection. His office is elbow-deep in papers, dog-eared notebooks, paint tubes, pastels, brushes, and clutter — seemingly unaltered since his death. Everything has a yellow cast. Malina was a smoker, and his tobacco still hangs in the air.

Near the entrance to the study, high on a wall, is a small, whimsical sketch of a cat. It is by the English caricaturist Ronald Searle, who was on the governing board of *Leonardo*. A bust of

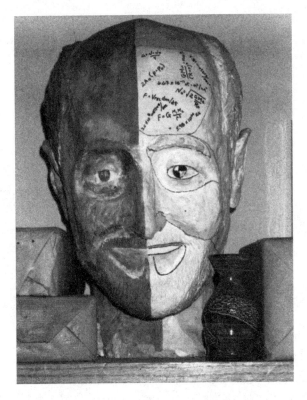

A bust of Frank Malina in his suburban Paris office.

Malina sits on a shelf behind his desk. Made by the son of Andrew Haley, Malina's Aerojet colleague, it dramatizes the two aspects of Malina's persona. The right hemisphere is brilliantly colored, slathered with intense red, blue, and orange paint. The left is white, cool, austere, and lettered with meticulous equations. Directly above his desk stands a statue of Don Quixote. With its mustache, goatee, and rail-thin figure, the Spanish knight is a physical doppelgänger of Malina. As well as a psychological doppelgänger: Both men were idealists, who spent their lives tilting at windmills.

Malina's widow still holds a representative collection of his artwork. His early nudes, delineated with string, were disappointing. But when he began dealing with content that riveted him—machines—the images came alive. Plugged in and activated, his kinetic sculptures transform their dim storeroom. Dust and cracking paint become star clusters and galaxies. When one piece, visibly timeworn, had the chutzpah to creak, Marjorie punched its frame, which quelled the noise. "He liked them silent," she explained to me.

I found the sketch for the screenplay in Malina's study. He had stored it in a folder marked "MGM," along with notes for a war movie called *Shadow of the Wing*, on which he had served as a technical adviser.

Despite Malina's Czech heritage, Kafka he wasn't. His first sentence—"This story is to be built on the present stage of rocketry as the foundation, with a superstructure of the dynamic social problems now existing"—is about as clunky as a sentence can be.

The story centers on Franklin Hamilton, a young, brilliant rocket scientist, who is "highly sexed," as is "common among great men," a detail one imagines Parsons contributing. Hamilton teaches physics at the "Institute of Science," whose head bears a resemblance to Clark Millikan, a man whose "mind" had been "bought, unconsciously, by money." And he leads a team of rocketry pioneers modeled on the JPL crowd: a Chinese character based on Tsien, a mystic patterned after Parsons, a mechanic resembling Forman.

As the plot thickens, or congeals, given the stiffness of the prose, "a wealthy aircraft manufacturer" in the image of Henry Ford (who was known to have Nazi sympathies) asks to see a demonstration of the team's rocket work. The mystic, however, objects. Involvement with the manufacturer, he learns in a premonition, will lead to harm. Unable to stop the demonstration, however, the mystic messes it up another way: killing himself in an explosion—a premonition by Parson of the way he would actually die.

But that was not the stunning plot turn. The greedy aircraft manufacturer, Hamilton discovers, has a nefarious agenda. He plans to fund the American scientists so he can sell their rocket breakthroughs to the Nazis. Somewhere, deep in their unconscious, Malina and Parsons had sensed what was ahead. In general rather than specific terms, they predicted their marginalization in history and, more horribly, the cold-war-era ascendancy of the Nazis.

Rather than hand their research to the Nazis, Hamilton and his team burn their papers. In real life, Malina and his team didn't have to destroy their own work; anti-Communists did it for them. And the Nazis, in Malina's lifetime, through their links to the American space program, were triumphant.

The historical record, however, does not end in Malina's lifetime. It is constantly revised, as new information comes to light. Much has changed since 1981. One generation's expedience sometimes proves to be the next generation's shame. Such was the case with Project Paperclip. By the early 1980s, when the space race had been won, the dirty secrets in the dossiers of the former Nazis at the Marshall Space Flight Center began to leak out. In 1984, Arthur Rudolph, the head of the Saturn V program, renounced his American citizenship rather than face a denaturalization hearing based on his war crimes. Far from being dimly aware of the monstrous conditions at the Mittelbau Dora concentration camp, Rudolph had created them. He had honed his management skills, the skills that realized the Saturn V, the rocket that launched the Apollo capsules, as the coordinator of slave labor at Mittelbau Dora.

Von Braun was not charged with anything after the secret Paperclip records became available. He was not around to face an indictment. He had died of cancer in 1977.

Malina doesn't appear to have been the type to have gloated. But had he lived a mere three years longer, he might have enjoyed the Nazi reversal, especially given the forces— antifascism and adherence to principle—that shaped his difficult life.

Of course, Malina was not entirely unrecognized. He collected a wall of international citations: the Order of Merit from the French Society for the Encouragement of Research and Invention, the C. M. Hickman Award from the American Rocket Society, even, in 1967, a citation in science from the U.S. House of Representatives. And on October 1, 1968, JPL director William Pickering invited him to speak at the lab's commemoration of the 1936 rocket test, along with other living members of the team.

Malina, alas, was not alive for the rocket test's fiftieth anniversary, for which JPL built the display of department-store mannequins that represent Malina and his colleagues. Curiously,

At JPL's annual open house, these mangy department-store dummies represent the JPL rocketry pioneers at their 1936 groundbreaking test.

the battered figures, I learned from a friend in JPL's public information office, have taken on a life of their own. Moved, one assumes, by human wiseacres, they often turn up not merely in their usual spot, the courtyard outside JPL's von Kármán auditorium, but in unexpected places, such as the conference table in the public information office.

As a metaphor, I like the idea that the figures refuse to be hidden, that they keep popping up despite efforts to tuck them away. It bodes well for the men they represent. With revisions in the historical record, the men, too, may someday appear in a truly surprising place: their rightful one.

Gender Parity, Part 1

FROM SCIENCE FICTION TO SCIENCE FACT

In the summer of 1964, my father got it into his head that we should, as the slogan went, see the U.S.A. in our Chevrolet. He charted a car trip from San Diego to Boston, with an eye toward making each stop educational for a soon-to-be fourth-grader. There would be geology lessons amid the striated spires of Bryce Canyon; history lessons on the battlefield where Sitting Bull vanquished General George A. Custer, and architecture lessons outside Mies van der Rohe's spare towers on Chicago's lakefront. I would experience the character-building hallmarks of a "back-East" summer—mosquitoes, blazing heat, and stifling humidity—as well as some treats: fireflies, my aunt's homemade ice cream, swimming in a freshwater lake. Mother packed gear for three weeks, which we stowed in the middle seat of our Dodge station wagon, behind my parents, who sat in front. I secured the far back for myself.

The trip, it is fair to say, opened my eyes to new vistas—but not the ones that my father intended. Concealed by heaps of luggage from my meddlesome parents, I was free to read all day, which I did, soaking up novels and short stories by Robert A. Heinlein—everything he had written except *Stranger in a Strange*

Land, his 1962 novel, which, because it satirizes religion and deals with sex, was banished to the adults-only section of our local library.

In Heinlein's stories I glimpsed the woman I wanted to become. This was not the world of Arthur C. Clarke, in which women played, at best, bit parts. Heinlein's women were educated and sexy. Men viewed them wolfishly, but nevertheless as peers, not helpmeets. For example, in the story "Let There Be Light," a male inventor awaits the arrival of a certain M. L. Martin, a "biochemist and ecologist, PDQ, XYZ, NRA, CIO— enough degrees for six men." When Martin shows up, however, the inventor is mortified. Hours earlier, he had attempted unsuccessfully to pick her up in a bar. The *M. L.*, Martin reveals, stands for Mary Louise.

In "Delilah and the Space-Rigger," a woman radio engineer joins the all-male crew on a space station under construction 22,300 miles above the Earth. To the shock of the boss who opposed her assignment, her presence improves productivity. He reverses his position, fully integrating women into his workforce. The woman engineer is named G. B. McNye, short for Gloria Brooks. So impressed was I with these Heinlein heroines, who used their initials, that in September, when I returned to school, Mary Grace became M. G.

Although their plots were easy to follow, the subtext of Heinlein's adult novels sometimes went over my head. I read *Starship Troopers*, for instance, without realizing that, as some critics suggest, the war it depicts between humans (who prioritize the individual) and bugs (who prioritize the collective) is a parable of the clash between capitalism and communism.

Still, any child could grasp the daring ideas in *Have Space-suit — Will Travel*. It centers on an eighteen-year-old regular guy named Kip who wants to visit the moon, which, through a fluke abduction by aliens, he manages to do. Further adventures follow, including saving the human race. His partner for these escapades, however, is not a male sidekick, Robin to his Batman or Bud Barclay to his Tom Swift. It is an eleven-year-old girl, Pee-Wee, who is smarter, tougher, and braver than Kip.

The book also features "the Mother Thing," an extraterrestrial of indeterminate gender, who provides the love, support, and cuddling that children traditionally get from their mothers. This blew me away. Here was a place where girls could outthink grown men and mothering was a job, not a biological destiny.

These were not the messages I picked up from my environment — or from what, for example, Wernher von Braun said in 1962 when, during a talk at Mississippi State College, he was asked whether NASA planned to fly female astronauts. Of course the agency would fly women, the former Nazi quipped; the men in charge were "reserving 110 pounds of payload for recreational equipment."

Heinlein's support for women in space was not confined to the printed page. In 1969, while serving as a commentator on the first moon landing with Arthur C. Clarke, Heinlein made network anchor Walter Cronkite sputter by suggesting that women should have been on the flight. His notion of the ideal woman astronaut, however, was so weird as to be patronizing — Peggy Fleming, a perky Olympic ice-skater.

One frustration of having lost my mother while I was a child

My mother at her college graduation, 1938.

is that I never got to ask her certain questions, such as: Why wasn't she more like these Heinlein women? Why, after getting her B.S. and teaching high-school chemistry, had she left graduate school and taken a job in the human resources department of a public utility? And why, after I was born, had she never gone back to work?

My mother's longtime friend Betty Jane Nolan had herself received a B.S. in chemistry from Newcomb College (now part of Tulane University) in 1943, five years after my mother earned hers at Ursuline College (now part of Loyola University). Betty met my mother at New Orleans Public Service, however, not in a chemistry lab. During World War II, Betty said, companies like Shell Oil and Firestone recruited women in her class for jobs around the country. But because travel was difficult during the war, she chose laboratory work in the New Orleans branch of a national paint manufacturer, which had a wartime contract for, among other things, "retreating compound," an olive-green substance used for waterproofing tents. Her job ended with the armistice. We had to let Miss Nolan go, wrote the male author of her backhanded recommendation letter, because of her outstanding performance. Her next position would have been production supervisor in the factory—not a "suitable job for a woman."

"Weren't you outraged?" I asked.

"That was just the way it was," she replied.

Yet in worlds other than the one I knew firsthand, women had, in fact, defied such oppressive conventions. These were not just characters in fiction but real women who, between the 1950s and the 1980s, ascended through the ranks at male-dominated institutions like JPL. One could start this story with Dorothy Lewis, Frank Malina's secretary, who stands in sharp contrast to women engineers and scientists at JPL today. In the 1940s, the only "technology" the men let Lewis touch was a switchboard. Or one could begin in the 1970s, when Donna Shirley came into her own—both as an engineer and as a catalyst for cultural change. Events at JPL, however, did not happen in a vacuum. They are best appreciated as part of the larger history of women and technology.

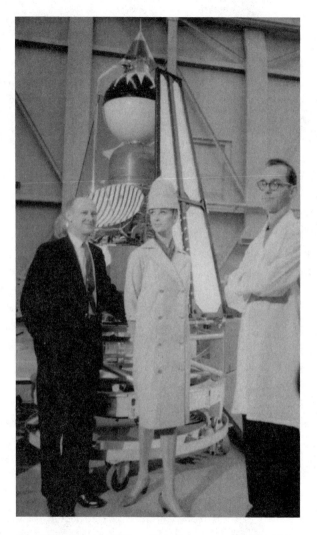

Then JPL director, William Pickering, poses with a fashion model and a technician in front of the Ranger spacecraft for a Harper's Bazaar *photo spread in 1962.*

✳   ✳   ✳

In *The Life and Death of a Satellite*, a nonfiction account of NASA's
robotic space program published in 1966, writer Alfred Bester,
better known as the author of such male-oriented, picaresque
science-fiction novels as *The Stars My Destination*, described the
role of women in the burgeoning space effort: They "are remarkably suited to delicate electronic techniques" but, after a
half-year on the job, "become bored by the niggling work, get
careless, and have to be fired."

The present-day reader, however, gets the sense that far from
being bored, women of the time should have been outraged—by,
among other things, a bizarre male fear of their biology. "Women
are not permitted to work on delicate components during their
menstrual periods," Bester continues in that book, without ques-
tioning the proscription. "Engineers dare not run the risk of
subjecting the components to the extra acidity of women's skin
at those times of the month." When I read Bester's reporting
to Bruce Murray, the director of JPL from 1976 to 1984, he
laughed out loud; it struck him as that scientifically unsound. In
1953, he added, when he began graduate school in geology, "the
conventional wisdom was that women shouldn't be allowed to do
fieldwork because they might be menstruating."

This irrational treatment of women was widespread. In the
early 1960s, for example, at Mount Wilson Observatory, just up
the Angeles Crest Highway from JPL, E. Margaret Burbridge, a
distinguished optical astronomer who would become a professor
of astronomy at the University of California, San Diego, was denied permission to use the telescope. The facility's "dormitory,"
where astronomers stayed during nighttime observation hours,

was restricted to men. Burbridge's husband, Geoffrey, however, was assigned time—never mind that, as Bruce Murray joked, he was a theoretician, whom "you wouldn't want to trust with the equipment"—and he turned the time over to his wife. Significantly, in 1972, the year Burbridge served as director of the Royal Greenwich Observatory in Sussex, England, she refused the American Astronomical Society's Annie Jump Cannon Award, which was intended to recognize outstanding accomplishments by women. It was inconsistent, she said, to fight for equal rights and accept a ghettoizing prize.

In *A World Without Women: The Christian Clerical Culture of Western Science*, published in 1992, historian David F. Noble makes a persuasive argument that today's masculine culture of science and technology evolved in the late medieval period, when access to higher learning was monopolized by celibate, ascetic, and frequently misogynistic priests. Such virtual eunuchs were repelled and terrified by women's biology—a horror they shared with the space-program engineers who banished menstruating women from the lab.

During the late nineteenth century, male anxiety about females in the workplace often took the form of paternalistic health concerns. In *Sex in Education*, published in 1873, Harvard physician Edward H. Clark argued that women's frail constitutions, and especially their fertility, would be destroyed by the strain of higher education. Yet other male scientists didn't even bother to cloak their fear of women. In 1880, for instance, two distinguished male chemists, Henry Morton, the president of Stevens Institute of Technology, and Thomas Sperry Hunt, a professor at MIT and fellow of the Royal Society, threw their colleagues a festive male-only gathering, which they christened the "Misogynist Dinner of the American Chemical Society." Not only were women excluded, the entertainment consisted of skits demeaning them. One recitation, titled "The Temptation of Saint Anthony," explicitly linked "the brave new men of science" with their "monastic heritage," Noble points out. "There are many devils that walk this world," the poem began. But "a laughing woman with two bright eyes *is the Worst Devil of All.*"

By contrast, the male colleagues of British scholar Mary Somerville (1790–1872), a self-taught master of mathematics and natural sciences, apparently held her in high esteem; they just had a strange way of showing it. Noble recounts that the men made her groundbreaking mathematical book, *Celestial Mechanism of the Heavens*, required reading at Cambridge University—a school that neither she nor any other woman was allowed to attend. And they placed a bronze bust of her in the Great Hall of the Royal Society, where neither she nor any other woman was permitted to set foot. (The Royal Society did, however, extend membership to her husband, despite the fact that he was not a scientist.)

"For many years there was little consciousness that these attitudes and practices might constitute something as ugly as discrimination," historian Margaret Rossiter writes in *Women Scientists in America: Before Affirmative Action, 1940–1972.* The first half of the twentieth century was not much better than the nineteenth. Nor, as one might have expected, did the drafting of men during World War II open doors to women. True, twenty-nine engineering schools that had not previously admitted women, including the Carnegie Institute of Technology, Columbia University's School of Engineering, and Rensselaer Polytechnic Institute, granted them access between 1941 and

1945. But this was a provisional solution, not a change in the cultural landscape.

"When the call went out for still more women engineers in 1942 and 1943," Rossiter explains, "civil engineer Elsie Eaves of the *Engineering News Record* warned women to be aware that recruiters used the word 'engineer' to describe two very different jobs: those few women with degrees in the field could expect beginning professional positions, but most others, including college graduates with a few additional courses in drafting and machine testing as well as women without degrees, were being used in subprofessional jobs as 'engineering aides' or temporary assistants to men who had recently been promoted from lower positions within the company."

Oddly, when they were founded in the nineteenth century, technical schools were intended as a less exclusionary alternative to the male-only seminaries of the Ivy League. Some actually fulfilled this promise. MIT, which was established as a land-grant college in 1863, enrolled its first woman student, Ellen S. Richard, in its second class. By 1895, an astonishing 6 percent of its students were women. Yet what happened at Rensselaer Polytechnic Institute belied the idealism of its pro-coeducation founder, Amos Eaton. In an 1824 letter, Eaton specified that his school be open to "sons and daughters," but until the midtwentieth century, the school only admitted sons.

Coeducation at Caltech was similarly stopped before it started. James A. B. Scherer, who in 1908 became the president of Caltech (then called the Throop Polytechnic Institute after its founder, Amos Throop), believed that were the school to admit women, few would matriculate—so few, in fact, "that it would not be just to themselves to come here." By masking his misogyny as concern, Scherer tabled the question of coeducation, without, as he put it, "exciting the militant suffragettes." As a consequence of Scherer's cleverness, Caltech managed to keep out women undergraduates until 1970.

*   *   *

Located in southern New Mexico, White Sands Proving Ground in the 1940s and 1950s was no place for a lady—or a woman, for that matter. For JPL engineers involved in developing the Sargeant and Corporal missiles, male-bonding rituals kicked off on the train from Los Angeles, with all-night poker games organized by JPL's deputy director, Louis Dunn. When they weren't at work, the men whooped it up across the border in Ciudad Juárez, Mexico. Such boys-will-be-boys antics were so integral to White Sands culture that they appear in a JPL-endorsed, midfifties documentary about the making of the missiles. The film's narration, which accompanies a picture of a rocket motor on a test stand, begins: "A conference was called to discuss the problem of how to get the motor started." It continues: "Positive action on two points was agreed upon. 1. Go to Juarez. 2. Catch Yvonne's late show."

Back in Pasadena, however, JPL's all-male environment was being swiftly coeducated—not with women engineers or scientists but with "computers," or "computresses." These were operators of Friden calculators, motor-driven mechanical devices that could add, subtract, multiply, and divide. Complex engineering feats, such as the computation of a spacecraft trajectory, required many such calculators working simultaneously. After the launch of Explorer 1 in 1958, spacecraft trajectories began to be JPL's stock-in-trade.

Introduced in 1952, the Friden SRW calculator weighed forty-two pounds, contained two registers and over one hundred keys. It was nearly always operated by a woman. An early advertisement for the firm, which was founded by Swedish immigrant Carl Friden in the 1930s, showed a voluptuous woman poised with her hands over the keys. Nor were the devices unique to engineering. At insurance companies, rooms full of women used them to compute actuarial tables. The relentless pounding, one listener remarked, was like the thrum of a marching army.

At JPL, computresses made up "Section 23," an all-female department that some engineers have compared to a convent and others to a harem. They were expected to have the devotion of nuns and to relinquish aspirations to the engineering priesthood. Insofar as JPL had a social season, it involved the competition among these women and other female staff for the title of Miss Guided Missile. Although a torpedo brassiere might thrust a contestant to the forefront—one campaign manager described his candidate as "a shapely craft, 5'6" in height, payload 120 lbs of well-designed equipment"—beauty alone would not secure the title. Aspirants had to mount the sort of popularity contest that one associates with class office in junior high. This was not a marginalized pageant; it dramatized the impunity with which JPL men objectified women. William Pickering, the director of JPL from 1954 to 1976, himself crowned the winner. In 1959, after the formation of NASA, when JPL turned its attention from missiles to planetary probes, the title became the Queen of Outer Space.

As technology evolved, however, the Friden seraglio became obsolete, replaced by the room-size IBM mainframe computer. (As did the Queen of Outer Space, which vanished in 1970 with no explanation.) Such devices were a far cry from the desktop models engineers use today; a single computation often required stacks of punch cards. But they were more streamlined than their forebears. The Mark I, for example, on which Harvard computing pioneer Howard Aiken began work in 1937, weighed five tons and was roughly half the length of a football field.

By the 1980s, computers had become much smaller, bringing powerful processors to individual desks. They also brought a fresh wave of women to JPL—newly graduated software engineers. Matt Landano, project manager on the Mars Odyssey orbiter, ascribes this influx to more than mere coincidence. Engineering, he believes, holds greater allure for women today than it did in the 1960s because of software. Computer engineering is a tidy, cerebral, indoor occupation. Engineers no longer "have to get grimy taking engines apart."

Although Landano's thesis is half-facetious, it describes a legitimate pattern. Educated forty years ago as mechanical engineers, Landano and his top-level contemporaries had focused on hardware. Confronted in the late 1980s and 1990s with puzzling new software, they were forced to rely on "fresh-outs," JPL slang for recent graduates, who were fluent in the latest computer languages, and many of whom were women.

"It was really weird as a 'fresh-out,'" JPL engineer Jan Berkeley recalled. "You're twenty-one, twenty-two, twenty-three years old, and you're working with people old enough to be your parents. The son of one guy I worked with did actually go to

JPL's Queen of Outer Space (formerly Miss Guided Missile) with her court.

school with me. But because you're dealing with a subject no-body else knows that much about, you're the expert."

Berkeley, who is African-American, is typical of the women engineers who arrived at the lab in the late 1980s and early 1990s. She grew up in Altadena, near JPL, and went to Pasa-dena's John Muir High School, but computers, not planets, drew her to engineering. "Whenever they showed computers on TV or in movies, there were big black boxes with blinking lights," she recalled. "I wanted to know what the lights did." She found out at the University of Southern California, from which she received a B.S. in electrical engineering in 1986.

In 1984, Berkeley spent a summer working at JPL. Her job was to translate a program that determined power allocations for the Galileo spacecraft from Fortran, the language "spoken" by JPL's mainframe computer, to Lotus 1–2–3, the language "spoken" by a PC. At the time, she recalled, there were only two PCs in her assigned section, and "not too many people knew how to use them." Hired by JPL in 1987, she was named the system fault protection engineer on Galileo in 1989, shortly before its October launch. While the spacecraft was on the ground, Berkeley's job involved finding strategies to deal with simulated problems. "Someone"—usually an engineer responsible for one of the spacecraft's systems—"would come up with a failure scenario, and we would tell them, okay, this is how we see this fault and how we would correct it." Each exercise took about a day, and the series of exercises went on for about nine months.

"Fault protection is like a parent. It keeps saying, 'Don't do that; don't do that; don't do that, and then you finally get punished.'" Or, if not exactly punished, sentenced to what Berkeley calls a quiescent state, known as *safing*, like the "safe mode," on a desktop computer, during which problems can be diagnosed.

Since its launch, Galileo has had a string of women fault protection engineers. In 1993, when Berkeley moved to the Cassini mission to Saturn, she was replaced first by Kauser Dar, who left to join Boeing in 1996, then by Z. Nagin Cox, who served as the deputy team chief on the Mars Exploration Rovers mission. Berkeley herself returned to Galileo in 1997, where she remained until 2003, when the mission ended.

* * *

When you ask JPL retirees about women who had a big impact on the laboratory in the 1960s and 1970s, they immediately mention two names: Marcia Neugebauer and Donna Shirley. Neugebauer, a physicist, was the first woman project scientist on a JPL mission. Shirley, an engineer who rose to head the Mars Exploration Program office, was a managerial reformer who questioned established practices. Where Neugebauer studied waves, you could say, Shirley made them.

The other key figure in these seventies-era transformations was Bruce Murray, the director of JPL from 1976 to 1984. In terms of two-career couples and women in the workplace, "the 1960s had passed JPL by—not just JPL but the entire aerospace industry," Murray said. His mission, among other things, was to connect JPL to changes in the larger culture.

Neugebauer joined JPL in 1956, when her husband, physicist and astronomer Gerry Neugebauer, was hired by Caltech. She had a B.S. in physics from Cornell University and an M.S., also in physics, from the University of Illinois, where she had studied the "anomalous scattering of energetic particles in nuclear emulsions." And not, she jokes, successfully. She got lots of data measuring microscopic particle tracks in a darkened closet. But the anomalous scattering, she learned later, had been caused by the closet's air conditioner cycling on and off. Nevertheless, as a consequence of her on-paper credentials, JPL assigned her to a sexy but short-lived project: the design of a nuclear-powered rocket.

Before the creation of NASA in 1958, JPL was run by the U.S. Army, which was in a heated missile race—officially against the Soviet Union, but mostly against the U.S. Air Force. "Rooting for the Army's Jupiter missile, we were sometimes treated to movies of spectacular failures of the Air Force's Thor missile," she recalled. Shortly after Neugebauer arrived, the U.S. secretary of defense, concerned that the interservice rivalry was out of hand, forbade the army from building a rocket that landed more than one hundred miles from its launch site. This prohibition nixed nuclear propulsion. The interservice rivalry also led to the formation of NASA, on December 3, 1958, as a *civilian* agency.

Neugebauer began studying ionized gasses, known as plasma. With a colleague, she detailed experiments that might be flown on the sort of planetary missions to which JPL aspired. She herself proposed to study the solar wind, a species of interplanetary plasma, then termed "solar corpuscular radiation," with a device called SCREPA—short for Solar Corpuscular Radiation Electrostatic Particle Analyzer. In 1962, her team was offered the opportunity to fly this instrument on two robotic spacecraft going to the moon: Rangers 1 and 2.

More significant, Neugebauer was named project scientist for these two Ranger missions. This was a major post, the link between Ranger's engineering team and its scientists. It was not, however, a post that lent itself to making friends.

As project scientist, Neugebauer had to divide limited resources between various scientific instruments, each of whose principal investigators, or PIs, believed his instrument deserved more of those resources than the others. The resources included funding and electrical power, as well as mass, which, in Earth's gravity, would be referred to as weight. (Like a jockey or a prize-fighter, the spacecraft had weight constraints; it had to carry certain instruments and still be light enough to launch.)

Soon she found herself, she recalled, in "the center of a three-sided battle" between the PI of the magnetometer experiment, the PI of an engineering experiment, and the project manager, who felt his mandate was to test the Ranger technology, not to gather science data.

As it happened, however, no one got what he wanted. Rangers 1 and 2 were notorious failures. They did not reach the moon. They remained stuck in the "parking orbits" they were supposed to demonstrate. (Parking orbits, the thinking went, expanded a spacecraft's daily firing window. You could send a spacecraft into low Earth orbit anytime, then fire a rocket to push it to its final destination—the moon, for example—when its alignment with that destination was ideal.)

Neugebauer picked her battles carefully, with an eye toward modifying, rather than destroying and remaking, existing management structures. When, within the newly formed space science division, she requested a promotion from "group leader" to "group supervisor," her boss said, "You don't earn enough money to be a group supervisor." She added, "I quickly suggested a cure for that."

She also helped amend JPL's policy on maternity leave—codified, she joked, by "the Caltech medical staff, who had never seen a pregnant woman before." In the 1950s, when Neugebauer had her first child, pregnant women were required to vanish without pay three months before their due date and not reappear until three months after. When a pregnant colleague on whom Neugebauer relied balked at leaving, Neugebauer challenged the rule, which was changed shortly thereafter.

Marcia Neugebauer, JPL's first woman project scientist, in the 1960s.

Neugebauer gathered data on the solar wind with instruments on such spacecraft as Mariner 2, which flew by Venus in 1962, and the Apollo Lander Science Experiment Package, which flew on Apollo missions in 1969 and 1971. She also helped design an ion mass spectrometer aimed at analyzing the mass and spectra of "fast or hot ions" that one would find, for example, in the coma of an active comet. A version of this flew in the 1980s on the European Space Agency's Giotto mission to Halley's comet.

By the 1990s, Neugebauer had become less of an anomaly at JPL as well as in the field of space physics. From 1991 to

1994, she served as chairman of the Committee on Solar and Space Physics, a subcommittee of the National Academy of Science's Space Studies Board, and worked as a study scientist in designing, among other projects, the Ulysses mission to the sun. She retired in 1998.

*　*　*

Although Neugebauer influenced changes in JPL's maternity policy, it took a man — Bruce Murray — to implement a child-care center. Or, more accurately, it took two men: Murray, who was then the director of JPL and his deputy director, retired lieutenant general Charles H. Terhune, a former fighter pilot and engineer who at Convair in the 1950s supervised the development of the first intercontinental ballistic missiles. (He was also an architect of Mutually Assured Destruction, or MAD, the cold-war defense strategy that ruled out limited nuclear war.)

In the 1970s, "women knew better than to bring child-care issues into the workplace," said Susan Foster, a JPL technical editor. Such issues "could be a reason for getting rid of them — if they were absent too much because their children were sick." Outside the aerospace industry, however, many businesses offered child care, and Murray believed it could aid in the recruitment of women and two-career couples.

In 1977, Murray created the Director's Advisory Council for Women (ACW), modeled on the Director's Advisory Council for Minority Affairs, which had coalesced under Pickering, and invited Neugebauer to lead the group. She declined. "I knew that if I chaired this committee, I would frighten everybody else away," she said. So Murray chose Kay Haines, a plucky technical librarian with a reputation for diplomacy, as its first leader. Foster would follow in October 1979.

Before JPL purchases spacecraft components from a contractor, it requires that contractor and others to submit a bid. This process ensures that JPL gets the best product at the best price. In contrast, Terhune and Murray did not make providers compete for JPL's child-care franchise. They invited a single candidate to meet with the lab's Executive Council: Eric Nelson, an unkempt, shaggy-haired, bearded nursery-school teacher, educated by Quaker pacifists. Before the interview, during what Nelson termed his "full-blown hippie days," he did not own a suit. He bought one, however, which proved to be a good investment. He was hired on the spot.

"They didn't know enough about early childhood education to be scared," said Elyssa Nelson, Eric's wife, who was also involved in the founding of JPL's Child Education Center. "And we didn't know enough about government bureaucracy and company life to be scared."

The center's first years were bumpy. It weathered a rat infestation and a forced closure by the L.A. County Health Department after an outbreak of giardia, an intestinal parasite. The parasite was linked to the center's wading pools, where the naked offspring of JPL's best and brightest regularly splashed. After the giardia episode, some parents tried to remove Nelson—first by voting him out, which failed, then through an audit by the NASA inspector general's office, which did not turn up fraud. Many credit Nelson's survival to the support of Donna Shirley, whose daughter was among the center's earliest clients.

Shirley also helped devise and diagram the center's organizational structure, a far cry from the top-down, hierarchical model one might expect at a NASA facility. At the center of a series of concentric circles is a heart, which represents the children. The heart is orbited by a moon labled *P,* for parents. Three levels of teachers surrounded the parents and children, followed, in an outer orbit, by administrators.

According to this model, demands aren't imposed upon the children from above, as they would be in a top-down arrangement. Rather, the staff responds to the demands of the children.

* * *

Shirley often says that once she "learned all the passwords" required to play with the boys, her goal was to change those passwords, as well as the clubhouse to which they granted admission. One skirmish at a time, she took on the Christian clerical culture of Western science.

Although Shirley acknowledges that some of its scholarship is flawed, Riane Eisler's *The Chalice and the Blade*, published in 1988, was a big influence on her. Eisler argues that some prehistoric societies functioned according to a "partnership model," in which no one — and, in particular, neither gender — seized a position above the other. This was replaced by Western society's present structure: the "dominator model," in which one gender asserts its primacy. The dominator model is symbolized by the "lethal power of the sharp blade" and is organized around the "slaughter of other human beings along with the destruction and looting of their property and the subjugation and exploitation of their persons." The partnership model, by contrast, is one of "linkage" and is symbolized by the chalice.

Eisler doesn't just slam the patriarchy. "The underlying problem is not men as a sex," Eisler writes, "but a social system in which the power of the blade is idealized—in which both men and women are taught to equate true masculinity with violence and dominance and to see men who do not conform to this idea as 'too soft' or 'too effeminate.'"

The book provides a road map for idealists to "intervene" in today's "cultural evolution," steering society away from the dominator model. As a map, the book's effectiveness is not weakened by the absence in history of any actual partnership societies. What is more, with the possible exception of the U.S. military, it's hard to imagine a more bedrock "dominator" than NASA. So Shirley, in trying to alter JPL, had to push a very large stone up a very steep hill.

Shirley believes in doing away with hierarchy and chucking the "tree" of linear management. Diagrammed, her ideal management structure resembles that of the Child Education Center: a series of concentric circles. At its center, instead of children, are the creative people. Managers and supervisors form the outermost orbits.

Eisler's thinking wasn't the only trendy, pop-culture influence on Shirley. Her unpublished book, "Managing Creativity," has a diverse best-seller bibliography, with works ranging from James Gleick's *Chaos: Making a New Science* (fractals, Shirley suggests, can provide ideas for management structures) to Michael Hammer's *Reengineering the Corporation: A Manifesto for Business Revolution* to Ray Bradbury's *Zen in the Art of Writing*. She even cites Uta Hagen's *Respect for Acting*—not entirely surprising, since much of her off-lab time is spent in amateur theatricals.

If a single incident could be said to have radicalized Shirley, it occurred in 1991, when she accepted the onerous task of leading a team assigned to diagnose NASA's management problems—not just at JPL but throughout the agency. Instead of producing a document that would ruffle no feathers, however, Shirley called it as she saw it, drafting what she called the "no-wimps" study.

"NASA is an entitlement program," Shirley summarized what she had written in the study. The job of each of its centers is to keep itself in business. "If you had to lie, cheat, and steal to sell projects to keep yourself in business, you would," she said. "And when you lie, cheat, and steal to sell a project—surprise!— you can't deliver what you said you'd be able to." NASA's "institutional objectives," she added, "are in conflict with its project objectives." This leads to a situation in which "no project manager on Earth" can deliver according to performance specifications because the project "has been underbid."

When Shirley presented these controversial findings to Rear Admiral Richard H. Truly, the NASA administrator, and other agency brass, she was attacked. But she held her ground. "I'm from JPL," she said. "I've been yelled at by experts."

The report, however, had little impact. It was released in a watered-down form, and none of its suggestions, to her knowledge, were implemented. To commemorate the episode, Shirley hung on her office wall a small photo of herself with Admiral Truly, mouths clenched in grins. It is near the room's most conspicuous object, a giant poster of Albert Einstein. The poster is emblazoned with a quotation that one cannot fail to view as a comment on the photo: "Great spirits have always encountered violent opposition from mediocre minds."

✳   ✳   ✳

Shirley's early years in Wynnewood, Oklahoma (population 2,500), were not characterized by strife. When she took flying lessons in high school, nobody disapproved—Amelia Earhart had made the image of androgynous aviatrix alluring, not threatening. Nor did Shirley question traditional femininity. In 1957, she was crowned Miss Wynnewood and proudly wore her sash.

During her freshman year at the University of Oklahoma, things took a limiting turn, but Shirley was not radicalized. Girls can't be engineers, her faculty adviser said, and instead of defying him, she graduated with a bachelor's degree in "professional writing." Only later, working in St. Louis as a technical writer for McDonnell Douglas, did she realize that in the aerospace field nonengineers were nonentities. She returned to Oklahoma for a bachelor's degree in engineering.

Freshly credentialed as an aerodynamicist, Shirley began work at Douglas in St. Louis on a Mars entry vehicle. Planetary exploration excited her, but opportunities were limited at a contractor. Determined to pursue her interest, Shirley joined JPL in 1966, where, by day, she worked as a mission designer for the Mariner 10 trip to Mercury and, by night, earned a master's degree in aerospace engineering from USC. The lab also had the fringe benefit, she joked, of being "a great place to meet men." In 1974, she married Tom Pivirotto, a JPL propulsion engineer.

Three years later, however, the birth of their daughter, Laura, was a watershed moment for her—and, like the "no-wimps" study, it had a radicalizing influence. She was stunned by the way men reacted to her becoming a mother. In public, her male managers refused to give her a "responsible" assignment. "They said, 'We thought Donna was now fulfilled as a wife and mother and just working for the fun of it,'" she recalled. (This may sound

Donna Shirley, 1973. The photograph accompanied a collection of "Profiles in Progress" (read: bios of women) in the March/April issue of JPL's in-house newspaper.

shocking in the late 1970s, but it is a reminder that, as Murray said, the 1960s had passed JPL by.)

What occurred in private was no less insidious. Although she and her husband had been "pretty equal," this ended with her maternity leave. "Since I was home, I was doing all the chores and I just kept doing them," she said. "He never picked them up again." This and other problems stressed the relationship, but Shirley struggled to make it work.

"I had this belief system that I couldn't take care of my daughter on my own," Shirley said. "I had allowed myself to be seduced by my mother's mind-set. I thought I wasn't strong, that I was dependent," perceptions, she realized, that were belied by her achievements. She and Pivirotto separated in 1983.

As Shirley's résumé shows, her career did not stall for long. She was soon assigned to manage increasingly pricey and complicated projects, from civil engineering work to the International Space Station. By 1991, the year of the "no-wimps" study, she was named chief engineer of a $1.6-billion project to explore the asteroids, comets, and Saturn. But Shirley's career—and her ascendancy in JPL management—was not solely the result of technical skill. Unlike many engineers, she could talk.

Beginning in 1974, Shirley served as a "voice," an official spokesperson for the lab who could explain the technical dimensions of a mission while the mission was unfolding. Moreover, Shirley was funny—a first-rate stand-up comic who could deliver the same line over and over without it growing stale. In the months preceding the launch of Pathfinder, for example, I heard her describe its off-the-shelf modem dozens of times: "Motorola told us, if we took it to Mars, it was off warranty." Her delivery always got laughs.

Al Hibbs, a physicist, JPL "voice," and key figure in the launch of Explorer 1, was among the first to spot her talent. In the 1960s, in addition to his JPL responsibilities, Hibbs was a television personality; he flew to New York once a week to host a Saturday-morning science show for kids. There are very few people, Hibbs said, "who could express themselves as well as Donna." Fewer still were also engineers.

As Shirley's star rose, some colleagues who commended her in public grumbled about her in private. Brian Muirhead, for example, the deputy project manager on the Pathfinder mission, had nothing but praise for Shirley in 1997. "Donna is way up on the totem pole, but she doesn't act like it," he said. "She has ideas—principles—about the way people ought to be treated. You don't see her managing upward—telling the boss what he wants to hear. You see her leading by example. She won't go home at five if she expects her team to be up all night working."

But in his book about the mission, *High Velocity Leadership*, published two years later, he paints a different picture. Shirley's appointment to the post of Mars program manager "was not a cause for celebration," he writes.

Ironically, considering the partnership model she espouses, some male managers faulted Shirley for being too dominating. She had, for example, locked horns with Tony Spear, the Pathfinder project manager, whom many had considered a "shoo-in" for the Mars program manager job. Besides the humiliation of having a subordinate named boss, Muirhead writes, Spear had to "overcome a personality conflict that had led to some fierce shouting matches." As Muirhead tells it, Shirley had also tried unsuccessfully to have Spear yanked from the mission. "Our best hope," Muirhead writes, "was that she would leave us alone."

The account is not out of keeping with Shirley's beliefs. Male managers—even younger, supposedly enlightened male managers—circle the wagons when one of their own is threatened. JPL "really is an old boys' club," Shirley said. "And the old boys are going to have to die off before it's anything but an old boys' club. But they're getting there—another five years, and they'll just about all be gone."

Two years after saying this, Shirley, too, was gone. In 1998, at age fifty-seven, Shirley took early retirement. Although she kept her reasons vague, the gag gift that she presented to JPL director Edward Stone at her going-away party spoke volumes. It was a copy of "Managing Creativity" with a special subtitle: "A Prophet Is Without Honor in Her Own Country."

Shirley may have felt scorned, but the skits at her party suggested that she had gained grudging respect. (When somebody is really in trouble, nobody dares to poke fun at them.) The program contained a lot of not-so-veiled aggression, beginning with its title: "The Real Donna Shirley Story as Told by the Men Who Wouldn't Listen." Her management ideas were also spoofed, in the form of a nonheirarchical organizational chart arranged in concentric circles—with Shirley at its center.

Nor did the men miss an opportunity to parody women in a general way. Roger Burke, Shirley's section manager in her early years at JPL, donned the Mars-red suit she had worn during Pathfinder press conferences and minced his way through an impersonation, which seemed less about Shirley and more about the aesthetics of drag. The attempt at metonymy—identifying a woman as a "skirt," a symbolic object associated with women—

was, with Shirley, off the mark. At the party as most days on the job, Shirley wore trousers.

But when the Cassini Virtual Singers, a group of singing "rocket scientists" attached to the Cassini mission to Saturn, crooned their parody tribute, I heard genuine admiration. Their lyrics paraphrased a Frank Sinatra classic: "She Did It *Her* Way."

* * *

On December 3, 1999, about a year after Shirley left, a mission called Deep Space 2 (DS 2) was supposed to land on Mars. At the time of the scheduled landing, the mission had a woman project manager (Sarah Gavit), a woman chief scientist (Sue Smrekar), and a woman lead engineer (Kari Lewis). The all-woman lineup was a coincidence, the result of a midproject departure by its male lead engineer, who was replaced by his assistant. But it did not go unnoticed.

DS 2 involved two basketball-size "penetrators," which had been programmed to plunge into the Martian soil and send back news about its water content. They flew to the Red Planet on the Mars Polar Lander spacecraft. And on December 3, when the Polar Lander failed, DS 2 went down with it.

I will never forget Gavit's stricken look that night on the stage before a crowd of journalists in von Kármán Auditorium, or the ashen faces next to her—those of Richard Cook, the Mars Polar Lander project manager, and Sam Thurman, its chief scientist. To the right of the dais, a life-size display of an intact lander on an ice-covered crater mocked their anguish. On Mars, the actual spacecraft was probably a heap of crushed metal.

"It was like a death," Gavit said.

For a time it looked as if Gavit might also have to mourn her career. She was not scapegoated or forced into industry. But neither was she immediately tapped to run a new mission, as male managers, linked to equally devastating losses, have been. Glenn Cunningham, for example, project manager on the Mars Observer, which was lost in 1993, swiftly took the helm of its successor, the Mars Global Surveyor.

Gavit's post-failure experience reveals a lot about the place of women in technology at the end of the twentieth century. No longer is the question, Will a woman be allowed to succeed? But rather, If she fails, can she recover and move on?

In the summer of 1999, six months before the scheduled landing, Gavit was confident, optimistic, at the top of her game. Models of the DS 2 aeroshell cluttered a shelf in her office. Some aeroshells resembled foot-long bullets, a design that was abandoned because of its instability. The final design was round and weighted like a badminton shuttlecock, whose heaviest part would hit the ground first. Because of financial constraints, the probes had a single-stage entry system, with no parachutes, rockets, airbags, or other devices to break their fall.

After slamming into Mars at over four hundred miles per hour, each probe was supposed to have separated into two parts: an aftbody, contained within the aeroshell, and a forebody, which would travel two feet deeper, releasing a tiny drill to collect a soil sample. The sample would then be pulled back into the aftbody and heated, to detect frozen water.

In 1995, Gavit's team contracted with a skydiving outfit in Mojave, California, near Edwards Air Force Base, to begin testing their designs. For over a year, skydivers hurled different DS

2 models out of a Cessna, and the team studied the way they crashed. The probes were subject to contradictory requirements. On the one hand, they had to be supersensitive, containing the delicate components of a laptop. On the other, they had to be ultratough, able to function after falling out of an airplane.

After the Polar Lander failed, instead of sympathizing with the JPL scientists, the country ridiculed them. "I had a very quiet weekend," Jay Leno told the *Tonight Show* audience on Monday, December 6. "The phone didn't ring. Nobody checked in. I felt like an engineer at NASA." He added, "NASA has a new game show. 'Who wants to lose $165 million?'"—an allusion to the price of the lander. That fall had been grim for JPL. In September, its Mars Climate Orbiter also failed. The lab and its contractor, Lockheed Martin, had crossed signals in an embarrassing way. JPL computed flight measurements on a metric scale; Lockheed Martin figured them according to the customary American system.

Perhaps because of its public drubbing, NASA was harsh in its internal review, filed in March 2000. The review team, chaired by Thomas Young, a retired executive at Lockheed Martin, was made up of eighteen technical experts, as well as six other consultants, including Bruce Murray, who had flown an experiment on DS 2. The report concluded that the DS 2 probes "were not adequately tested" and "were not ready for launch." It also accused Gavit of being "inexperienced."

Gavit's team, the report said, had not tested the probes according to a JPL rule: "Fly as you test, test as you fly." This means you must prove that your hardware works by subjecting it to the same nasty conditions it will have to endure in flight; then, after launch, stick to the plan you have tested. Because a fully functioning DS 2 probe had not been shown to survive a four-hundred-mile-an-hour impact on Earth, Young's commission questioned whether it could have survived one on Mars.

"In hindsight," Gavit observed, "I can't disagree with them saying that we should have done the test. We should have. In an ideal world, that's what we all would have liked to have done. But the real world is not ideal. We did not have either the money or the time in our schedule to build an entire third spacecraft."

The charge of inexperience also stung. Her inexperience — which implied a willingness to resist hidebound approaches — was precisely why she had been named to the post, according to Kane Casani, a former manager of the New Millennium program, of which DS 2 was a part. "The New Millennium program was supposed to break with tradition," he said.

Gavit spent much of calendar year 2000 drafting a formal response to the Young report. Her emotional response, however, poured out in a talk she gave at one of JPL's afternoon "story hours," as some engineers call them. These lectures take place in the laboratory's library and, atypically, emphasize feelings as well as facts.

Gavit's talk was a long poem inspired by Dr. Seuss's *Green Eggs and Ham.* In her version, a beleaguered "Sarah Ann" begs a NASA authority figure to send her "Minicam" into space. Beginning with size and deadline restrictions, the NASA mucketymuck makes impossible demands, just as NASA's "better, faster, cheaper" directive had placed nearly insurmountable limitations on DS 2.

"But if it were just half its weight / One third its size, it can't be late. / Then I might like it, Sarah Ann! / I just might like your Minicam," the muckety-muck says. Soon "Sarah Ann" has

Sarah Gavit, 2004. In 2003 Gavit left the Dawn mission to work as a manager on the Jupiter Icy Moons Orbiter (JIMO).

agreed to push the envelope—to a point beyond which the envelope cannot be pushed.

"I started drafting the poem in my car at lunch," Gavit recalled. "The next thing I knew, it was five at night." Catharsis, however, was not enough. The experience made her question her professional choices, swept her into a maelstrom of uncertainty. First as a student at MIT, then as a young engineer at Lockheed Martin, she had aimed her whole life toward a career at JPL. She stayed in the car. She deliberated. "Do I still feel the passion?" The answer, when she dared to look, came plainly and promptly: "I do. Some people have the bug in them to explore. And I'm one."

In March 2000, Gavit was involved in planning a mission to test the feasibility of a "solar sail," a gadget that would harness the solar wind to propel a spacecraft. NASA, however, decided not to fund the mission. If this dismayed her, she didn't say. Gavit is a committed engineer. And the goal of engineering is to make things work even—or perhaps especially—with constraints on the design.

By 2002, Gavit's dedication had paid off. She was named project manager on a new mission that will launch in 2006. Its spacecraft will travel to the main asteroid belt, between Mars and Jupiter, where its focus will be on Vesta and Ceres, two large rocks viewed by some as dead baby planets.

The mission's name has resonance—both for Gavit and for other women at the turn of the millennium. Considering the dark night that for Gavit preceded it, the mission, appropriately, is called Dawn.

No Lost Opportunity
or
THE LAUNCH OF MER-B

Nearly every JPL engineer I met urged me to witness a launch from Cape Canaveral Air Force Station. To comprehend the work that they did—the enormous power of a rocket, the heft and delicacy of a spacecraft—I needed to watch that spacecraft leave the Earth.

My father felt that way. After a few fingers of Scotch, the preferred tipple of his sunset years, he would speak reverently about the idea of a launch, as if it were Easter Mass at St. Peter's Basilica or a private audience with the pope. Between the end of our estrangement and the point at which he was too weak to travel, we talked about going to a launch together. He investigated public viewing areas for the shuttle. But other commitments interfered, and we never made the trip.

One can easily interpret a launch as a symbol of masculine power. It involves a potent object penetrating the heavens. But what, I suspect, gives the experience its psychic resonance is that its imagery is also feminine. During preparations for a launch, the spacecraft is "mated" to the launch vehicle. The connections between the components are "umbilical." The harsh, male abstractions of physics are described in the vocabulary of pregnancy and birth, the ultimate female acts of creation. A launch is

A Delta II Heavy rocket blasts off from Cape Canaveral carrying the Opportunity rover, July 2003.

more than just a knock-your-socks-off explosion. It is a kind of labor, a wrenching struggle to escape one world and move into another. On a deep, archetypal level it is profound—a uniting of the masculine and feminine in the sort of balance that Jungian psychologists believe the psyche seeks.

Or so I conjectured. I needed to witness a launch—or, more specifically, to examine a launch as a work of theater—to see if my speculations were borne out. And I regretted not having made that trip with my father. I wanted to see a launch for him.

By 2002, many of the people whom I'd gotten to know on the Galileo project in 1997 had moved to a sexy mission called Mars Exploration Rovers, or MER, that would land two golf-cart-size rovers on the Red Planet in 2004. The rovers were to be launched in the summer of 2003, and I resolved to watch at least one lift off. Each was to launch separately, during a period that began in early June and extended to the middle of July. I could not attend the launch of the first rover, MER-A, nicknamed "Spirit," which took place June 10. But the launch of the second rover, MER-B, dubbed "Opportunity," fit my schedule. It was set for 12:38 a.m. on Wednesday, June 25.

Four years earlier, in June 1999, I had witnessed, or, to be more accurate, *felt* a launch from Vandenberg Air Force Base. But the experience was so truncated that I couldn't make sense of it. I stood on a grassy hill with reporters and photographers, awaiting the ascent of NASA's Quick Scatterometer, a satellite whose principal instrument was designed to collect data on winds near the ocean surface. The instrument was produced by JPL; the satellite, by Ball Aerospace Corp., part of the Boulder, Colorado, empire that makes Ball jars for home canning. Vandenberg is located on an unspoiled strip of California coast near

Santa Barbara, and in the early evening, when the rocket was set to go, a thick fog had rolled in. What I mostly remember are smells — sage, which grew in clumps along the bluff, and burned coffee, poured from a thermos in the trunk of a photographer's rented Chevy. The coffee, along with some hard, greasy doughnuts, seemed startlingly delicious as we huddled against the cold, wet air. QuikSCAT was set to launch on an enormous rocket, the Titan II, a recycled Titan intercontinental ballistic missile, so I expected major fireworks. And at about 7:15, when the rocket fired, the ground did indeed shudder. I had to plug my ears against the roar. But because of the fog, I saw nothing. Not even a spark.

"Did the Earth move for you, too?" a guy working for the contractor asked.

Well, yes. But I had expected more.

I thought of that risqué joke as I few from L.A. to Florida for the launch of MER-B. It struck me as yet more evidence that launches bring to mind sex, that quintessential mingling of male and female. I had cleared my schedule that week to allow for weather delays and technical glitches, which proved to be wise, since the launch was postponed, from the wee hours of Wednesday morning to a little before midnight the following Saturday, June 28.

This was due to a technical glitch. Because MER-B was being sent up near the end of the optimum period for launch, it required a rocket with greater oomph than the Boeing Delta II that had launched MER-A. This rocket was a Delta II Heavy, which has six solid rocket boosters strapped to its first stage. (The first stage itself burns liquid fuel.) At the flight readiness review, engineers determined that a band of cork insulation on this stage was not adhering properly and would have to be reglued. The band is about two feet wide and a quarter inch thick — a reminder that little things can be critically important.

When I picked up *Florida Today*, the local Gannett paper, I understood the MER team's caution. The paper was running a seven-part series on troubles at NASA. It highlighted the agency's decision to launch the space shuttle Columbia in January 2003 with sixteen hundred known problems that could have destroyed the spacecraft and killed its crew. Under scrutiny for recklessness, NASA had to show prudence.

NASA has a big stake in the way the public views its programs. Cold-war-era engineers, however, often disagree with their younger counterparts about how its image should be managed. Engineers from the 1960s have a sense of entitlement; they remember the Race, when money flowed without seeming limit, and "rocket scientists" were aloof and lionized. Younger engineers, by contrast, are used to implementing the "better, faster, cheaper" directive of former NASA administrator Dan Goldin. They understand that funding is affected by the whims of Congress, which are, in turn, affected by the whims of the electorate.

Although NASA is prohibited from self-promotion by the 1958 Space Act, it courts goodwill in many ways, ranging from theme parks associated with its regional centers to collaboration with sympathetic moviemakers. In the 1950s Walt Disney and his adviser, Wernher von Braun, fueled interest in space travel through the Rocket-to-the-Moon ride at Disneyland and the 1955 "Man in Space" and "Man and the Moon" episodes of Disney's TV show, a tradition that continues to this day.

So Sunday night, stranded for a week in oppressively sticky, sun-soaked, mosquito-ridden Florida, I planned to check out NASA's public relations presence on the so-called Space Coast. This meant a trip to the visitor complex at the John F. Kennedy Space Center, located in the middle of a vast, seabird- and alligator-filled nature preserve, and a day at Disney's Epcot Center, located an hour to the west of Cape Canaveral in Orlando.

NASA has a long history with the entertainment industry. In 1998, buoyed by the public-relations triumph of the 1995 movie *Apollo Thirteen*, NASA headquarters in Washington, D.C., installed Bobby Faye Ferguson as its first full-time liaison with the movie industry. Ferguson, a former TV actress who had had a recurring part as a hooker on *Designing Women*, was an unlikely choice, but sources say she was a friend of President Bill Clinton, and the position was a White House appointment. The idea of such a position is not new. For more than fifty years, the Pentagon has maintained a link with moviemakers who want to use military installations or equipment, and the policy has had definite benefits. According to Philip Strub, the Pentagon's Hollywood liaison in 1998, the 1984 hit movie *Top Gun* reversed some antimilitary sentiment caused by the Vietnam War—a crucial feat, when one is building a volunteer army. Ferguson was also midwife to flattering portrayals of her agency, such as HBO's 1999 series *From the Earth to the Moon*, a multipart chronicle of the Apollo program. And during her tenure, NASA firmed up a contract with IMAX to document the building of the International Space Station.

In addition to the Kennedy Space Center, NASA's other flagship theme park is Space Center Houston, a 183,000-square-foot visitor complex that is a joint project of the NASA Johnson Space Center and the not-for-profit Manned Space Flight Education Foundation. This Texas destination offers exhibits, souvenirs, and a tram tour of JSC.

In contrast, JPL has had a modest approach to image management. Its visitor's center is a small room — seventy-two feet long, thirty-six feet wide, and thirteen feet tall — with a ragtag collection of spacecraft models, and, on special occasions, that mangy crèche of department-store dummies. JPL's educational outreach plan, however, is extensive. Through its Solar System Ambassadors program, for example, ordinary citizens compete to be emissaries for JPL projects; those who qualify are invited to JPL for training.

This is not to say that JPL hasn't aspired to a flashier presence. In April 1997, its public affairs director, George Alexander, commissioned a firm that designs theme parks to present a feasibility study for a new visitor complex, including a motion ride that would whisk passengers on what Alexander termed "a magic carpet through the solar system." He pictured visitors driving on the surface of Mars, plunging into the ice-covered ocean on Jupiter's moon Europa, and watching the birth of stars. While no new visitor complex emerged from the meeting, the JPL displays, which in 1997 had a sixties high-school-science-fair feel, have at least been replaced by exhibits with updated graphics.

Still, when it comes to the relationship of NASA and the entertainment industry, there is such a thing as too close. Mike O'Neal, a member of the team that designed the air-bag landing configuration for the Mars Pathfinder, may be the emblem of the resourceful, younger engineer at odds with his hidebound, older colleagues. In the early 1990s, while working full-time at JPL and studying film in UCLA's extension program, he wrote a paper for JPL suggesting that NASA look to Hollywood for new ways to finance and manage its projects. Making a film, he argued, was a lot like carrying out a space mission. Both are big-ticket, one-of-a-kind items produced by teams, which come into existence with budgetary and schedule constraints. In the 1990s, the cost of a "better, faster, cheaper" space mission was similar to that of a big Hollywood feature: about $150 million. (By contrast, MER, a mission with two spacecraft, cost more than $800 million.)

O'Neal also recommended that missions be supported by advertising. How much, he wondered, would a company's logo on a Mars rover be worth to that company? Enough to pay for the rover? "Let Pepsi be a proud sponsor of a mission, like it was a proud sponsor of the Olympics," O'Neal told me. "Did that cheapen the Olympics? No, it enabled the Olympics."

O'Neal's suggestions, however, fell on deaf ears, due in part to the disdain with which older engineers received them. Placement of a corporate logo on a spacecraft would violate the 1958 Space Act, which precludes such advertising, and Congress was unlikely to amend that act without a groundswell of support. Frustrated, O'Neal left JPL for the industry he had investigated, joining Digital Domain, a Venice, California, digital special-effects house. This move highlights another similarity between missions and movies: The technical skills necessary to land a spacecraft on a planet are close to the ones required to create the illusion of that landing.

O'Neal's departure is part of a brain drain that was brought to national attention after the Columbia accident. In February 2003, the *Los Angeles Times* reported that NASA has had trouble attracting and keeping younger engineers and scientists. Employees over sixty outnumber those under thirty by about three to one.

Even after the dot-com bust, NASA could not compete with private industry on salaries. In March 2003, NASA administrator Sean O'Keefe testified in a Senate hearing that for NASA to hold on to top employees, it must have better hiring and retention bonuses, as well as an increase in its highest salary, from $134,000 to $142,000.

The numbers paint a depressing picture, but a brighter one reveals the passion of younger engineers who choose space work. They don't do it for the money but because, as Sarah Gavit put it, they have the "bug" to explore. Seven people I interviewed in the 1990s have since left JPL. By 2003, one had returned, one is trying to return, and two would like to return, although they are not actively seeking a job change. Even O'Neal said that he would be pleased to go back, if the right project came along.

* * *

The Kennedy Space Center is a museum of the analog era, a celebration of sixties hardware, a place where the space shuttle, viewed by many post-*Columbia*-disaster critics as a seventies relic, is palmed off as cutting-edge. It is also a museum of cold-war masculine values.

To board the bus for the deluxe tour was to enter a time machine. Our bus driver, a thirty-year KSC veteran, told stories about the Apollo astronauts as if they were still frisky boys chasing waitresses around nearby Cocoa Beach. He invoked the name of Jim Rathman, the car dealer who let them obtain Corvettes for close to nothing, and dwelt lovingly on their life-threatening driving. By the time the bus disgorged at its final destination, the Firing Room Theater, a mock-up of the ground-based nerve center for the Saturn V launch that bore up Apollo 8, the last three decades might as well never have happened.

The space center would have met my father's standards for an edifying trip. So well did it evoke the mid-1960s that I felt as if he were there. I pictured my mother, too—with a cocked eyebrow supporting a mission named for a pagan god.

The dramatized launch was, to be sure, gripping. Audience members stood around a reconstruction of the firing room, made from original consoles, status boards, countdown clocks, and communications gear. The equipment looked primitive and creaky, like that smelly old space helmet to which I had been perversely attached. No people were shown, but one would be hard-pressed to imagine a more masculine environment—from the emotionless "Roger-A-OK" conversations between astronauts and controllers to the gray, utilitarian furniture to the 363-foot, 6.2-million-pound, relentlessly phallic Saturn V, offstage yet at the center of the drama. Given the crudeness of the technology and the magnitude of the journey, one also has to admire everyone's bravery.

Apollo 8 launched just before Christmas in 1968—a year of intense national pain, most especially from the assassinations of Dr. Martin Luther King in April and Senator Robert F. Kennedy in June. On Christmas Eve, while in orbit around the moon, the astronauts read aloud from the book of Genesis. My parents and I were moved when we heard the words, "And God said, Let there be light," as a lunar sunrise washed over the capsule. But that was not the only affecting end-of-the-year news. We were also stirred—and disturbed—by the increasing body count in Vietnam.

When school resumed in January 1969, my social studies class discussed the events of the holidays. One boy pointed out what he viewed as a paradox: Here we had this amazing ship that could travel 240,000 miles to the moon. But it could not traverse a bigger divide—the rift in the country over Vietnam.

In many ways, KSC seemed to belong to a different universe than that of JPL, not a place where exploration is "human" or "robotic," but rather, as our bus driver put it, "manned" or "unmanned," those noninclusive terms of the 1960s. Even the center's tip-of-the-hat to JPL, a timeline of the robotic spacecraft that it produced, is backhanded. The coded-for-cuteness automatons who narrate the exhibit characterize themselves as mere "trailblazers for human exploration," as if the glimpses of new worlds that Mariner or Voyager or Galileo sent back were not in themselves impressive.

So backward-looking was the whole experience that the center seemed moribund. When I saw turkey buzzards flying around the Vehicle Assembly Building, a 526-foot-tall structure that dominates KSC, I could not help but think of the vultures that circle dying creatures in animated cartoons.

Mission: SPACE, by contrast, the seductive new virtual-reality ride at Disney's Epcot center, could persuade even Luddites to support a trip to Mars. It does not allude to a triumph of the past. It excites riders about triumphs yet to come.

Wisely, the team that created it—Hewlett Packard, Disney's Imagineers, and some NASA advisers—invented their own space agency, the International Space Training Center (ISTC), to support the fictive mission. They link this agency to NASA's groundbreaking firsts through a photo gallery of the real-life

Mission: SPACE—a simulated Mars flight that opened at Walt Disney World in 2003.

astronauts who achieved them. But they also distance this agency from NASA's less-palatable legacy, such as those sixteen hundred known flaws on the space shuttle Columbia.

The ride, reported to have cost in excess of $100 million

(about two-thirds the cost of the Mars Pathfinder mission), is beautifully and lavishly produced—from the mother-of-pearl paint on its curving exterior to the shimmering orange and blue spheres in front of it. Not portrayed to scale, these spheres represent Mars and the moon, and look like giant, glitter-flecked bowling balls. Inspirational, heart-in-your-throat *Chariots of Fire* music floods the foyer; a multiroom maze conceals the length of the line.

At first, some kids I observed seemed put off by the ride's educational content, and I wondered if they were on vacations with instruction-minded parents. Quotations from such textbook figures as Galileo Galilei ("The universe stands continually open to our gaze") and astronomer Carl Sagan ("The surface of the earth is the shore of the cosmic ocean") adorn the exterior walls. An authentic Apollo lunar dune buggy, lent by the National Air and Space Museum, is mounted in a room inside. But when Gary Sinise, who played astronaut Ken Mattingly in *Apollo Thirteen*, appeared on video as the mission's Cap-Com, or flight director, the kids perked up. He warned that the ride was scary. People who don't like high speeds or small, dark enclosures would do well to get off.

I am not a roller-coaster fan, but I boarded the four-person module and strapped in. As the simulated rocket engines fired, we were smashed back into our seats. I felt my skin stretch into a too-tight facelift. Objects flew past on the flat-plasma computer screens in front of us, creating the illusion that we were hurtling forward. Then, in simulated orbit, something more amazing happened. We felt weightless—a sensation with which I am familiar, having experienced it two years earlier on the KC-135, the so-called Vomit Comet, a NASA cargo plane adapted to fly in parabolas.

Mission: SPACE does not end in Earth orbit. The four-minute Mars trip features more rapid-fire surprises, based on new technologies NASA is studying. The space agency, a spokesperson said, was pleased to have advised Disney on the ride because of the opportunity to "inspire the next generation of scientists, engineers, technologists, and astronauts." And, of course, taxpayers. Why else select as its slogan, "We choose to go"?

✳ ✳ ✳

At Cape Canaveral, launches are highly ritualized, with an almost sacramental sequence of events. The buildup to MER-B was no exception.

It began Friday with the liturgy of the word, which took the form of a press conference. Then, later in the day, on the launchpad, observers gathered for a benediction: the roll-away of the gantrylike support structure that surrounded the rocket. Performed before a congregation of video cameras, this unveiling was powerful. The rocket went from its Clark Kent concealed identity to 128 feet of glistening, muscular Super-vehicle.

The ceremony helped me understand cargo cults. Worship seems a reasonable response to a giant, achingly beautiful object that hums with mystery and danger. And it harked back to the linkage of Western science to Christian monasticism. Viewed in this context, for example, the weird vestimentary rituals practiced by Apollo mission director Gene Krantz make a sort of sense. For each flight on which he served, Krantz wore a specially made vest—similar to the color-coded chasubles that priests wear for specific feasts.

Launches also unfold in hyperprecise, stylized language, vastly different from conversational speech. For ten days before the MER-B launch, NASA sent out press releases in this language: "On June 17, the MER-B rover, Opportunity, contained within its payload transporter, rolled out of the Payload Hazardous Servicing Facility at 1:58 a.m. It arrived at Pad 17-B at 4:30 a.m. and was hoisted atop a Boeing MER-B Delta II Heavy Launch vehicle at 9:00 a.m."

The process seemed so steeped in left-brain, masculine communication patterns that I was surprised to see a woman, Kris Walsh, the director for NASA programs at Boeing Expendable Launch Systems, seated onstage at the press conference. So fluent, however, was she in affectless, technical speech, that her gender was soon unnoticeable. She explained an aspect of the Delta II Heavy that would heighten its drama at launch—a "sound suppression system" to guard against acoustic energy that might damage the spacecraft. This system is essentially a big tank of water dumped into the flame pit beneath the rocket just before blastoff. As a consequence, during the first seconds after launch, observers don't see the rocket ascend. They see clouds of steam.

The rocket's second stage, she told us, like the second stage of MER-A, was equipped with forward- and aft-facing cameras. (When MER-A launched, the aft-facing camera sent back striking pictures of Florida vanishing in the distance.)

From a theatrical standpoint, only Steve Squyres, a Cornell astronomy professor and the principal investigator for the rover instruments, would have looked out of place at a sixties press conference. He wore a black shirt, black Gap jeans, and cowboy boots, appearing, if not fully of-the-moment and hip-hop, at least post-cold-war and rock-'n'-roll.

Cornell University astronomy professor, principal investigator for MER rover instruments, and rock-'n'-roll scientist Steve Squyres.

After alluding to "tantalizing clues" about Martian water that his team planned to investigate, Squyres walked over to a life-size model of the mission's "robotic geologist," a six-wheeled, 358-pound rover whose outstretched solar panels made it look like a winged dinosaur. The gadget's "head," supported by a stovepipe "neck," featured two "eyes," high-resolution color stereo cameras. It also had multiple spectrometers and a cunning jointed forearm that held a device to grind rocks. Bigger and tougher than the *Sojourner* rover, it was nevertheless cute, though its body type bore an unfortunate resemblance to the raptors that slaughtered tourists in *Jurassic Park.*

In December, Squyres had told a reporter that the rover's cameras would send back images clear enough to "look good projected on an IMAX screen," a description that finally made sense. IMAX, I learned, was making a movie about MER, with Squyres at its center. In his ability to explain and excite people about science, Squyres reminded me of Carl Sagan, the telegenic astronomer, also from Cornell, who was associated with JPL and gained recognition as a television personality and author.

As Squyres was clearly the man to watch, a friend and I accepted an invitation from some Cornell University News Service people to attend a launch-night party where he was the featured attraction. The party was for Cornell alumni. It would take place somewhere on the waterfront in Cocoa Beach. They would ring with details.

* * *

The following night, however, when our hosts failed to ring, we struck out for Trident Bluff, the brush-covered viewing area reserved for the press and named for the Trident submarine that sometimes docks below it. Launch time was set for 11:56:16 p.m. A second time was available at 12:37:59 a.m.

By 10:30, photographers and TV crews had gathered in a parking lot outside KSC—as had clouds of ravenous mosquitoes. Blow-dried television reporters halted their field reports to shoo bugs. I doused my clothes, shoes, and hair with acrid-smelling insect repellent and prepared for the worst. At 11:15, just before our official NASA escort was set to shepherd us into the Air Force Station, the cell phone rang: our Cornell contact. Finally. With the location of Squyres and his entourage.

We left KSC and drove to Jetty Park, where hundreds of launch-watchers milled or picnicked near the bustling bait-and-tackle shop. There were teenagers on dates, families pushing strollers, children in T-shirts bearing pictures of the rover against the pink Martian sky. Near the ocean, there were blessedly few mosquitoes. A podium with a microphone was set up outside the bait-and-tackle shop. Speakers from the space community, including Bill Nye, the so-called Science Guy, an engineer turned children's broadcaster who had designed a sundial for the Mars rover, addressed the crowd. If Vandenberg had smelled of sage and KSC of bug spray, Jetty Park reeked of beer. "I think this is what they used to call a Happening," my friend observed.

Squyres and the Cornell group were midway out on the jetty. Crowds clustered at the jetty's tip, where the rocket could be seen without obstruction—a majestic white obelisk glistening above the polished black water.

Near Squyres, a Cornell news photographer set up his camera. He planned to capture not only the rocket's fiery trail but its reflection in the water.

The IMAX crew, however, had something more ambitious in mind. They planned to interview Squyres while the rocket whooshed upward, and to show a path of warm light across his face, as if he were illuminated by the moving rocket. As rockets can be fickle, the effect would have to be achieved artificially. The crew attached a pink light to the end of a boom, wrapped fabric over it, and practiced sweeping it across the grassy dune where Squyres would be planted.

Over a crackling loudspeaker, we heard the launch readiness poll, an exchange reminiscent of a responsorial psalm. The mission director ran through a checklist of systems on the spacecraft. Then, if the system was working properly, the person in charge of it said, "Go for launch."

At about 11:30, a patrol boat motored by — odd, since the launch area was supposed to have been cleared. But we thought nothing of it until T minus four and counting, when the clock was placed on hold.

As it happened, the boat should not have been there. The coast guard turned up a fisherman who had strayed into forbidden waters. The countdown had to be stopped. Almost immediately, the MER team began reprogramming the rocket for the second firing time.

As theater, these developments were both annoying and suspenseful. Squyres spent the downtime on his cell phone. Glowing with self-satisfaction, the IMAX team practiced its thousand-dollar-a-minute shot, caressing the faces of some cooperative bystanders with rosy light.

When the clock approached T minus four for the second time, tension heightened.

Again we heard: "Go for launch."

The water was quiet. No breeze stirred the palms overhead.

"Go for launch."

Heads swiveled toward the rocket.

"Go for launch."

Photographers pressed against the eyepieces of their cameras.

"Go for launch."

Then, suddenly, nothing.

At a far higher altitude than we could perceive, the wind had picked up—to a speed that might imperil the rocket. Launch was postponed until Sunday, the following night.

By Sunday afternoon, it was postponed until Wednesday. By Wednesday, it was moved to Saturday. The cork was acting up again.

I lifted the handset of the hotel phone to arrange another week in central Florida, but my hand froze. I realized that I had seen what I had needed to see: the sacramental buildup to the launch, the tumescent Delta, throbbing against the indigo sky. These were the masculine components of the picture, which had evolved in the 1960s and remained unchanged. The feminine components were veiled, embedded in language, not pagentry; the "umbilical" connection between the spacecraft and the rocket, for example, was not put on display.

For a more balanced picture—if there were such a thing—I needed to see what was happening at JPL.

*   *   *

When I entered von Kármán auditorium to join the "friends and family" event for the launch, I knew that I had been right to leave Florida. The stage held video screens of varying sizes, displaying feeds from the launchpad, the KSC firing room, and the JPL control room. The KSC monitor could have been playing a tape from the 1960s; it showed wall-to-wall white men. The feed from JPL, by contrast, put on view faces that were of this century and female. As I made my way up the aisle, larger-than-life images of Z. Nagin Cox, MER's deputy engineering team chief, and Grace Tan-Wang, deputy chief of its "sequence" team, which prepares commands to be sent to the rovers, flickered on-screen.

Earlier in the evening, at 5 p.m., the rocket had been filled with refined kerosene; at 5:51, with supercold liquid oxygen—a process that, when completed, caused its bluish airframe to appear dazzling white.

Charles Kohlhase saved me a place near the front of the auditorium. As manager of science and mission design for the Cassini-Huygens voyage to Saturn, he had sat in the firing room at Cape Canaveral on October 15, 1997, when that spacecraft launched. I asked him to meet me at JPL to hear his impressions. He began by explaining that "liftoff" was not a vague term, but a specific one, indicating that the rocket had moved two inches upward from the pad. "Place a mark on the rocket anywhere you like, and when it has risen two inches relative to the launch platform, you have liftoff."

Because of the time difference between the coasts, the launch would happen in Pasadena about 8 p.m. As a consequence, von Kármán was full of kids. They scrambled after posters, guest badges, and buttons as if they were swag thrown from a Mardi Gras float.

Jan Ludwinski, the master of ceremonies, invited the children to the front of the room. There, within feet of the screens, they squirmed with anticipation, expressing all the hope and eagerness that the grown-ups felt but were too dignified to convey.

The flight readiness poll had begun. With the patience of Mr. Rogers, Ludwinski fielded questions from the children. They asked the sort of things that I had asked Kohlhase. Why does a spacecraft launch to the east? To take advantage of the Earth's rotation. What do "TECO" and "MECO" mean on the diagram of the spacecraft's trajectory? "TECO" is an acronym for third-engine cutoff; "MECO," main engine cutoff. They refer to the point at which a stage of a rocket burns out.

Ludwinski had honed his master-of-ceremonies expertise at other "friends and family" events. In December 1996, he kept order during Galileo's first flyby of Europa by leading the auditorium in a Jupiter-themed version of the word game Pictionary.

"Stand by for release of hold," a loudspeaker broadcast. But nothing happened.

The countdown was halted. There was a problem, Ludwinski announced, with a "fill-and-drain valve" on the Delta II first stage. It would have to be examined and tested.

No sooner had the first countdown been scrapped than the team began preparing for the next. They put the errant valve through repeated tests. If all went well, a countdown to the second firing time would resume at T minus four minutes.

The children groaned. This was, after all, a school night, and as it grew later, they were more likely to be banished to bed.

I remembered the room full of inner children at Donna Shirley's workshop and wondered what wounds these sunny youngsters would carry into adulthood. I thought of my own father, who had been so proud of Mariner 69 and ached to be linked with the bold men who flew it. I thought of Bill Nye, whose father had inspired the sundials on the rovers. Not only were these tiny devices, emblazoned with "Two Worlds, One Sun," terrific for public relations, they also performed useful chores. Because it reflects the sky, the gray background of each sundial appears pink on Mars; and the degree to which it appears pink aids scientists in determining the true colors of the Martian landscape. Nye first learned about sundials from his father, who studied them in a Japanese prison camp during World War II, where he had to discern the position of the sun to tell time.

I also thought of Claudia Alexander, the last of seven project managers on the Galileo mission, who would in September 2003 direct the spacecraft to crash into Jupiter. When its propellant ran out and its course could no longer be controlled, it had to be destroyed, lest it accidentally collide with one of Jupiter's moons and contaminate that moon with organisms carried from Earth. (Although Jupiter is viewed as unlikely to support life, its moons, such as Europa, with its ice-covered ocean, are considered hospitable, and if a NASA craft detects life, NASA wants to make sure it is not life imported from back home.) To many people involved with its fourteen-year voyage, the crash was a death. They had guided Galileo through many crises—a jammed high-gain antenna, a stuck tape recorder—and they felt parental toward it. But at an age when a human child would be preparing for the SAT, its "life" ended.

Alexander's other duties made it easier to part with Galileo. One day a week, she served as project manager on the scaled-down Galileo mission, and four days as project scientist on the U.S. portion of the nine-year European Rosetta mission to an asteroid. During the cold war, there were long lulls between giant missions like Galileo. Today, smaller robotic missions from JPL and other space centers launch frequently throughout a given year. One can view them not so much as individual entities but as parts of an ongoing quest for knowledge.

This change in point of view is, of course, rooted in economics. Better, faster, cheaper missions tend to be small, and to launch more often, without much attention from the general press. But I think the change also has to do with the end of the cold war. Maybe not at KSC, but certainly at JPL, one gets the sense that science missions are actually about science—not pretexts to dazzle enemy nations with technical prowess.

In an official video that played on launch night, the MER brass presented itself with introspection and humility. Their relationship to the spacecraft was nurturing and parental. "I'd be lying if I didn't say I was scared most of the time," said Jim Erickson, MER mission manager, who, like Alexander, served as one of seven project managers on Galileo. Richard Cook, who sat ashen in front of von Kármán when the Mars Polar Lander crashed, was again ruddy and chipper; he confessed, however, that the accident had plunged him into the "depths of despair for about two years." And project manager Pete Theisinger was a model of thoroughness and caution. "I worry about the thing not found," he said.

In 1997, about the time of Donna Shirley's workshop, I read *Reflections on Gender and Science,* a 1985 book examining the historic conflation of the so-called scientific with the so-called masculine, by MIT professor Evelyn Fox Keller. I underlined this passage: "What is called scientific receives extra validation from the cultural preference for what is called masculine, and conversely, what is called feminine—be it a branch of knowledge, a way of thinking, or woman herself—becomes further devalued by its exclusion from the special social and intellectual value placed on science and the model science provides for all intellectual endeavors."

With Cox and Tan-Wang looming large on the live feeds, Ludwinski impersonating a nursery school teacher, and Erickson, Cook, and Theisinger affecting soft, confessional tones, the feminine did not seem severely excluded.

The countdown restarted. As the numbers got lower, Ludwinski and the children, still going strong on a school night, shouted them out. "Five, four—"

The acoustic suppressors were due to release their torrent of water.

"Three—"

Steam exploded near the bottom of the rocket.

"Two—"

The clouds billowed higher.

"One—"

Loudspeakers filled the room with fierce rumbling.

A startled baby wailed.

"We have liftoff," Ludwinski said.

The children shrieked. Some knelt frozen, transfixed by the giant screen. Others sprang up, jumping, as if they were rockets themselves.

The Delta II roared skyward.

About three minutes later, the solid rocket motors burned out and were jettisoned. The first stage continued to burn, then fell away.

The second stage ignited—a light ball in the darkness. The rocket's Earth-facing camera sent back pictures, but they showed little. Only in daylight could one watch Florida disappear.

Well out of eyeshot, the third stage fired. By this time, the rocket had been boosted out of Earth orbit and was on its course to Mars.

Not until 9:43 West-coast time, however, did the launch officially end. JPL fight controllers received a signal from the spacecraft via the Deep Space network. The cruise phase had begun. The flight team cheered, as did what was left of the audience.

I walked outside to the parking lot, zipping my jacket against the cool summer night. Mars blazed overhead, radiant and ginger-colored. In less than a month it would appear bigger and brighter, the nearest it had been to the Earth in sixty thousand years. I shivered. The full import of what had just happened hit me. The spacecraft, whose violent departure I had witnessed, would in six months touch down on that shimmering pinprick of light.

Gender Parity, Part 2

FROM SCIENCE FACT TO SCIENCE FICTION

In 1969, Ursula K. Le Guin's The Left Hand of Darkness *won sci-*
ence fiction's most sought-after prizes, the Hugo and Nebula
awards. It was not your everyday space yarn, filled with swashbuckling astronauts and nifty hardware. It was a critique of po-
larized gender roles. I first read the book in 1970, not long after
my mother's death. But even had my grief not been fresh and
sharp, I would have been moved by its tragic love story.

Le Guin set her book on Gethen, a planet populated not
with men and women but with "potentials," beings that, for a
few days each month, become male or female. Most of the
time, however, they are hermaphroditic neuters, who can both
bear children and sire them. The book's central character is a
human man visiting Gethen, whose fixed gender and unwaver-
ing sexual preference troubles the natives, just as an ever-
changing gender and fluid sexual preference might trouble
some people on Earth. On Gethen, human men and women —
that is to say, people in permanent hormonal imbalance — are
called "perverts."

Against the odds, in the course of a gripping flight across
Gethen's ice- and snow-covered world, the human man and a
Gethenian "manwoman" fall in love, sacrificing their selfish hap-

Ursula K. Le Guin's haunting challenge to conventional gender roles was published in 1969.

piness for a goal that will benefit both their civilizations. The tale is a sort of extraterrestrial *Casablanca*, with an ending just as sad. As so often happens in life, the bravest, noblest character suffers horrible abuse and is killed.

Nor did only deep, moving books critique middle-class sex roles. Loopy, hallucinogenic, satirical sci-fi did, too. Set in a colony on Mars, for example, Philip K. Dick's *The Three Stigmata of Palmer Eldritch*, published four years earlier, hinges on a drug that allows people stuck in a godforsaken outpost to project themselves into an Eden of consumer goods. The drug-induced projection, which includes sexual experiences in idealized female and male bodies, involves elaborate miniature play sets and dolls modeled on Barbie and Ken.

For most of my adult life, my love of cold-war sci-fi — and tendency to hold forth boorishly about it — has not been a social asset. But when I started reporting on JPL, it proved to be a cross-cultural Esperanto. I had few life experiences in common with older male engineers; sci-fi brought us together. They had also escaped into Heinlein, Clarke, Bradbury, Le Guin, and Dick.

The feeling of closeness, however, also produced a puzzlement: How could people who claimed to have loved these books have been unaffected by their daring ideas — specifically, their challenges to traditional gender identities and sexual preferences? From the 1950s to the 1970s, the aerospace industry was a Noah's Ark of go-to-the-office husbands and stay-at-home wives. There were, of course, a handful of married women engineers, but those who did not fit the Noah's Ark paradigm in a more fundamental way — gay engineers and scientists, for example — were invisible, or, if they dared reveal their identities, persecuted.

Where open-mindedness was a hallmark of science fiction, bigotry defined science fact. In recent years, however, this has become less true, which may mean that gender-bending sci-fi as well as, of course, changes in the larger culture, have had some effect.

After World War II, persecution of gay people in the sciences intensified—claiming, among its casualties, Alan Turing, the distinguished English mathematician and computer scientist, who was openly gay. One could, in fact, argue that World War II could not have been won without Turing, who cracked the Nazi's notorious Enigma Code. He also built the Turing Machine, an early computer, and devised the Turing Test, a set of questions designed to determine if the respondent was a person or an artificial intelligence program.

In April 1953, President Dwight Eisenhower signed into law Executive Order 10450, which, for the first time in civil service law, stated that "sexual perversion" was grounds for firing or not hiring federal workers. Great Britain and the United States were closely allied, and as a consequence of the thinking that led to this order, Turing was denied a security clearance. In 1954, unable to continue his research, and commanded by an English court to undergo chemical castration (consensual sex between men was against English law), Turing took his own life at age forty-two.

Turing, of course, did not work at JPL or in the American aerospace industry. But his experience exemplifies the past bias against gay people in technical fields. During the space race, the West was willing to overlook Nazi war crimes to gain the upper hand. But not homosexuality.

In part, this is because homosexuality was conflated with the period's other great evil: communism. Like Communist cells, the

Alan Turing, the British computer scientist who cracked the Nazi's Enigma code.

gay community of the 1960s was concealed. It occupied the shadows, threatening, in the view of some, to subvert American institutions—a perception heightened in later years by the reve-lation that Soviet agents Guy Burgess and Anthony Blunt were, in fact, gay. Mainstream media such as *Life* magazine fanned the embers of this paranoia. In July 1964, *Life* ran a leering, voyeuristic photo-essay on "Homosexuality in America." "For every obvious homosexual, there are probably nine nearly impossible to detect," the magazine warned, where the word *detect* recalls J. Edgar Hoover's portrait of Communist agents in his book, *Masters of Deceit*. The magazine continued: "Do the homosexuals, like the Communists, intend to bury us?"

Oddly, *Life* ignored the fact that the Soviet Union repudiated and oppressed gay people, too. Just as the United States had stigmatized homosexuality as Communist, the Communists branded it as capitalist—a bourgeois aberration, the placing of individual desire over the welfare of the state. In 1917, the Russian Revolution overturned czarist sodomy laws, but by 1933, the Soviet Union enacted a new law that made consensual sex between men a crime, punishable by up to five years of hard labor. Homosexuality was considered a "treasonous desire," explained sociologist Laurie Essig in *Queer in Russia*, which "good citizens," who were "always straight," were obliged to "control, punish and eventually eliminate."

Where the Soviets tended to incarcerate gay engineers and scientists in psychiatric hospitals, the United States simply derailed their careers—by withholding security clearances. After Executive Order 10450, defense workers had to declare their sexuality when they applied for a job.

"In the sixties, if you were found out to be gay, [your employer] would look at this form you had to fill out when you were hired," said Gordon Wood, a former JPL deep-space communications expert who retired in the 1990s. "There was a homosexual tendencies box and if you'd checked it, they'd say, 'We never should have hired you.' If you didn't check it, they'd fire you for lying."

As a result, Wood did not draw attention to his affectional preference. A clean-cut electrical engineer who graduated from the University of California, Berkeley, in 1964, he obtained a job at JPL on Mariner Mars 69. He had grown up in nearby Glendale. He was a local boy, a boy-next-door. And he made sure that was how his colleagues perceived him—at the start of his career, anyway.

"I can't begin to tell you what you have to do as a male to pretend you're straight," Wood said. "When a young, attractive woman walks by, you have to sort of follow like this with your eyes"—he demonstrated a tracking leer—"like men do all the time. I would have to train myself in these traits that did not come naturally, to prove that I was like all the other guys sitting at the lunch table. I couldn't really let my eyes follow the people I wanted to look at."

Until the 1970s, security clearances were flatly denied to gay applicants. By the 1980s, however, the process had become murkier. If the Department of Defense determined that a gay candidate would not be subject to blackmail, the candidate might obtain a clearance. During Ronald Reagan's presidency, when funding for many pure science projects dried up, JPL competed for defense work, requiring many employees who had not previously needed security clearances to apply for them. These included Wood, for whom the slow, exasperating process came to a head when a DOD investigator turned up at JPL.

Sequestered with the investigator in a tiny interview room,

Wood said, "I came here to be an engineer and not a gay activist. But I don't mind who knows that I'm gay."

"Well, we're not so sure you don't mind," the investigator said, "because you don't tell everybody."

But since he first arrived at JPL, Wood had become more open. To demonstrate this, he left the room and invited a handful of milling security officers and human resources staff to wedge inside. He then announced, "I would like to tell you that I am a homosexual. Here is my mother's phone number and my sister's phone number. You may call them and discuss anything you like with them. This will not be a surprise to them." To the investigator, he said, "I don't think I'm subject to blackmail here," an assessment with which the inspector presumably concurred. Wood's clearance came through shortly thereafter.

By the early 1990s, however, such solo battles gave way to a group struggle, intended to revise social institutions rather than create loopholes for individuals. In the larger culture, the AIDS crisis had galvanized many politically indifferent people to identify as gay and resist discrimination. The slogan of ACT-UP, an activist organization formed at the time, was "Silence = Death." Against this backdrop, a small group of young engineers and scientists took the conflict to JPL.

They brought three key attributes to the struggle: They were open about their sexuality, so no one could level a charge of "double life"; they understood how to use media to expose and embarrass bullies; and they could pick themselves up after being knocked down. Cate Heneghan, a JPL technical staff member, and Annmarie Eldering, an environmental engineer who is now at JPL but was then in graduate school at Caltech, organized this group's first action.

In August 1989, Heneghan sought permission from JPL's human resources department to hold meetings of a gay and lesbian support group on the laboratory grounds. Alcoholics Anonymous and Weight Watchers met there; she expected no opposition. But Robert Sutherland, the head of human resources, refused, instructing the group to "go through Caltech to accomplish" their objectives.

So she went through Caltech, organizing a panel discussion with two campus organizations on "Homophobia in the Sciences." But when she sought to announce the event in the *Universe*, the JPL in-house newsletter, she was rebuffed. "This subject is not of general interest to the broader JPL community," a public information officer wrote.

Promoted only by word of mouth and campus posters, the panel took place January 30, 1990. It attracted more than 125 people, packing the conference room. It also attracted the *Los Angeles Times*, which published a story about high-tech gays, comparing Caltech and JPL unfavorably with schools such as UCLA, which had nondiscrimination language in its charter. In 1990, companies such as the RAND Corporation, Xerox Corporation, and Apple Computer, as well as universities such as Stanford and MIT, had antidiscrimination policies and authorized support groups.

In 1992, Caltech president Thomas Everhart finally amended JPL's policy to ban discrimination based on sexual orientation. But the battle continued for domestic partner recognition. In 1991, Lotus Development Corporation initiated health benefits for same-sex domestic partners—paving the way, by 2004, for hundreds of other major U.S. firms, ranging from Intel Corporation to the Walt Disney Company.

Seeking to implement such benefits, Heneghan persuaded JPL's deputy director, Larry Dumas, to meet with her group. The younger, outspoken members believed that Dumas should also hear from older, less vocal, more distinguished scientists and engineers—people like Wood and his domestic partner, Glenn Cunningham, who served as the project manager on the Mars Global Surveyor and the Mars Observer.

In the middle 1960s, when Wood and Cunningham came to JPL, they lived in different apartments within the same building. But so closeted and discreet was the cold-war gay community that they didn't become friends until 1977, when, during the launch of Voyager, they stayed in the same condominium complex in Cocoa Beach.

As partners, they faced problems common to accomplished heterosexual couples. When, for instance, the Mars Observer spacecraft failed, JPL sought Wood, because of his knowledge of deep-space communications, for its internal failure review board. Management, Wood said, was aware of their relationship and discussed the matter with them before naming Wood to the post. Conflict of interest was ruled out, Wood said, because the board's charter was not to place blame on anyone but simply to determine what had gone wrong.

Although Wood had proved his openness to the DOD investigator, he was still far from an activist. But, he recalled, "When Larry Dumas agreed to meet with a group of gay people at JPL to discuss domestic partner benefits, I said to myself and I said to Glenn, 'It's important that we stand up and be counted now. Not just for us, but for everybody who works here—and for future employees.'"

The meeting took place in the fall of 1993, just after Mars

Observer was lost. Turnout was high. And prestigious. There were senior managers, nationally known scientists, recipients of NASA distinguished service awards. Engineers who had lost partners to AIDS protested the lab's policy on bereavement leave. Heneghan pressed for the benefits, which were implemented within two years.

Today, the *Universe* announces meetings of the support group (which welcomes bisexual and transgendered people). Members can also stay in touch through an authorized e-mail list. Communication is open, simple, and no big deal—a status that, at the time of Turing's death, would have seemed like science fiction.

* * *

By October 2003, Opportunity was halfway to Mars, with its twin, Spirit, barreling along a few weeks ahead.

Scientists described the rovers' mission as "following the water," where *water* was a code word for life. For sci-fi fans, however, life, in one fictive form or another, has existed on Mars for decades, as well as on more distant worlds. Arthur C. Clarke's Nebula and Hugo award-winning novel, *Rendezvous with Rama* (1973), which details an encounter with an alien civilization, has particular relevance to JPL—not because it involves one of the lab's robotic explorers but because it is the first of a tetralogy whose last three books were coauthored with Gentry Lee, currently the chief engineer for all JPL flight projects.

Lee is one of those polymaths who, even at JPL, stands out. Like Clarke and Heinlein, Lee received a technical education, including a master's degree in aerospace engineering from MIT and a Marshall Fellowship to study mathematics at the University of

Glasgow in Scotland. But as an undergraduate at the University of Texas at Austin, where his father taught journalism, he studied languages and literature.

In 1976, he served as the director of science analysis and mission planning for the Viking mission to Mars (while working for Martin Marietta, which built the project's two landers), and later joined JPL as the chief engineer on the Galileo project. His current job is "to ensure the engineering integrity" of missions that travel outside Earth orbit. In that capacity, he ran engineering reviews of key parts of the MER mission, such as entry, descent, and landing, to make sure the team was prepared for the unexpected.

Besides collaborating with Clarke, Lee has published sci-fi on his own. He also codeveloped public television's *Cosmos* series with Cornell astronomer Carl Sagan. What he does best as a writer is engineer cliff-hangers; his plots are elegant and suspenseful. What he does less well is enter the heads of his characters, who, regardless of gender or background, tend to speak like midcentury male engineers. This has not stopped him, however, from writing in characters' voices. Lee opens *The Garden of Rama*, for example, the third book in the Rama series, with the journal of a woman astronaut, Nicole des Jardins.

Des Jardins is far from a slouch. Besides being a cosmonaut, she is a medical doctor, an Olympic athlete, and the breathtakingly beautiful biracial daughter of a West African princess and a French novelist. In her forties, without collapsing, she produces five children by two men, one right after the other. And even for a single parent, her concerns about her oldest child are exceptional. "Is it fair," she wonders, "for me to keep from my daughter the fact that her father is the King of England?"

But des Jardins does have one problem: She's too perfect. Women can't relate to her, and men are afraid of her. What she's missing is her inner PeeWee, the goofy, preteen sidekick in Heinlein's *Have Spacesuit — Will Travel*, an insecure inner figure that few mature women ever fully banish, and whose presence in a woman character increases that character's plausibility.

Lee's male characters, by contrast, are not implausibly perfect. Richard Wakefield, des Jardins's husband, is a shy, geeky guy who feels "more comfortable with the world of mathematics and engineering than he is with other people." Michael O'Toole, her fellow astronaut, must reconcile his devout Catholicism with the imperative, while in alien captivity, to propagate the species. The only person with whom he can procreate is des Jardins, and he does so, despite her marriage to Wakefield.

When I first met Lee in 2001, I asked if des Jardins was his ideal woman.

He said no, she was his ideal daughter, adding that in real life he has no daughters, but rather seven sons—several of whom were the result of his and his wife's strenuous efforts to produce a daughter.

I sympathized with Lee's disappointment, but I was also relieved. At least no teenage girl would have to suffer comparison with des Jardins.

When I saw Lee during the MER mission, however, his awareness of adolescent girls—and, by extension, adult women— had undergone a sea change. I wanted him to talk about how this mission was different from Viking. He wanted to praise Jennifer Harris Trosper, the mission manager for Spirit, and Z. Nagin Cox, the deputy chief of MER's engineering team.

Then, as if he had discovered an astonishing new fact, he said,

Martin Marietta engineer Gentry Lee during the Viking Mission to Mars.

"Our society absolutely discourages women from going into science and engineering."

You're kidding, I was tempted to sneer. Instead I simply said, "In college?"

"At the university level, no. Young women are encouraged at MIT and Caltech. The problem is between the ages of ten and sixteen."

His sons, he reminded me, are in middle school or beyond. "And I've had the good fortune to talk to some of their female acquaintances. These girls are very bright. I'll ask them the question, 'Why aren't you taking physics?' Or, 'Why aren't you taking calculus?' And they'll say, 'Because it's sort of nerdy.' Or, 'It doesn't go with the image of being in the popular crew.'"

This is where "articulate, fantastic role models" like Trosper and Cox come in, he said. You want young women to think, "I wanna be like Jennifer Trosper. Or I wanna be like Sally Ride. Not—pardon me—I wanna be like Paris Hilton."

Lee did make a few interesting points about the differences between MER and Viking, most of which involved computers. On Viking, the IBM mainframe would take all night to run a single trajectory; on MER, an ordinary laptop could run thousands of trajectories almost instantaneously. But I barely heard these differences amid his concern for adolescent girls.

Since all engineers in this story are on some level stand-ins for my father, this was a watershed moment, or, in any event, a moment of wish fulfillment. Lee said what I fantasized my father would one day say. For me, this was the point at which the cramped, closed compartment of science fact opened onto the expansive vista of science fiction.

✳ ✳ ✳

In January 2000, 405 of JPL's 2,548 employees in science and engineering were women. (The total number of employees in all areas was 4,646.) By 2004, Caltech decided that its ratio of men to women was "proprietary information," and kept it secret. A source close to the numbers, however, placed the amount of women on the lab's technical side at about 18.5 percent. This stands in sharp contrast to the 1960s and 1970s, when, throughout NASA, women made up only 2 percent of the scientific and engineering workforce—but 92 percent of the clerical staff.

The current number of women at JPL may seem insubstantial, but in terms of the way women scientists and engineers are treated in less-welcoming parts of the world, it is significant. Z. Nagin Cox, MER's deputy team chief, pointed this out in 2001, just after the events of September 11.

"Western cultures unilaterally treat women the best," she said. "If you look at what happened with women in the 1960s and '70s, we're talking about problems at a high level. Whether or not there's equal pay for equal work; whether or not there's respect in the workplace; whether or not somebody is telling offensive jokes—these are very advanced things compared with whether or not somebody canes you because your shoes tap," she said, alluding to a widely reported Taliban atrocity that we had been discussing.

Cox was born in Bangalore, India, to a conservative Muslim mother and an agnostic father. After a stint in Malaysia, her family moved to Kansas City, where her father, a university professor, taught political science. As she was growing up from age seven in the Midwest, Cox recalled, her parents were so bitterly divided on religion that they had provided no "spiritual baseline." So she slaked her spiritual craving through the mythological resonances and moral instruction of *Star Trek*.

That TV series also awakened her interest in real-life space exploration, a career that her father opposed for women. At age seventeen, in defiance of him, she became an American citizen. This enabled her to join the U.S. Air Force Reserved Officers Training Corps (ROTC), which would pay her tuition to study engineering. In 1986, having earned a B.S. in operations research engineering and a B.A. in psychology from Cornell University, she was commissioned as a lieutenant in the air force. While working as a systems engineer at Wright-Patterson Air Force Base in Ohio, she earned a masters degree in spaceflight operations from the Air Force Institute of Technology. Soon she had achieved the rank of captain and secured a plum post: orbital analyst at NORAD/Space Command deep in Cheyenne Mountain, Wyoming. Her next step, in 1993, was to leave active duty and join JPL.

When Cox first struck out for Cornell, her father was far from pleased. But as she rose in her profession, he came around. Even during the rough times, she recalls, the problems she faced were at a high level, of the sort encountered by women in a culture that treats women well.

* * *

In December 2003, as Spirit and Opportunity entered the final stretch of their journey, Donna Shirley was finalizing a journey of her own. Its destination was announced in February 2004.

The journey, you could say, began at age eleven, when Shirley read Arthur C. Clarke's *The Sands of Mars* and set her sights on a job in planetary exploration. After retiring from JPL in 1998, Shirley worked as a freelance speaker and consultant for a year, then returned to the University of Oklahoma—first as an

Donna Shirley (left) crouches near a model of the Sojourner rover, a half-scale version of which she donated for display to the Science Fiction Museum and Hall of Fame.

assistant dean in the College of Engineering, then as a professor of aerospace engineering.

Meanwhile, her daughter, Laura, entered graduate school in behavioral neuroscience in Washington state and got married. The ceremony harked back sweetly to Laura's toddlerhood. It was presided over by Eric and Elyssa Nelson, the founders of JPL's Child Education Center, who had been ordained as ministers in a nontraditional sect.

Shirley was not unhappy in Oklahoma. But she wanted to be near her daughter. In 2003, Paul Allen, the cofounder of Microsoft, gave her the opportunity to do so. He established a new museum in Seattle dedicated to science fiction and asked Shirley to direct it.

"Science fiction is metaphor," Ursula K. Le Guin wrote in a 1976 introduction to *The Left Hand of Darkness*. "What sets it apart from older forms of fiction seems to be its use of new metaphors, drawn from certain great dominants of our contemporary life — science, all the sciences, and technology, and the relativistic and historical outlook among them."

Who better than Shirley to curate such metaphors? On day one of her management course, she had insisted that her students master figurative language. And in a nice twist on the idea of an eleven-year-old girl inspired by an elder statesman, Arthur C. Clarke, a member of the museum's advisory board, told the press that he "could not think of anyone better" than Shirley to serve as its director.

Bouncing Toward Meridiani

or

POSTCARDS FROM THE NEW WORLD

Just before nine in the morning on January 24, 2004, I headed to a press briefing in von Kármán Auditorium. The San Gabriel Mountains behind JPL were crisp and snowcapped, different from the shrouded forms I recalled from Shirley's inner-child workshop. Gray clouds still obscured bits of them. But the eye-stinging smog was gone.

The Spirit rover had touched down successfully on Mars on January 3, making front-page headlines. Shirley, drawing on her experience as a JPL "voice," had provided play-by-play commentary on its landing at a Planetary Society gathering in the Pasadena Convention Center. The rover continued to grab attention: In the three weeks since it reached Mars, NASA's Web site received over 4 million hits, more than double the number to the Web site in all of 2003.

Several days earlier, however, while lumbering to a rock that scientists had named Adirondack, Spirit had stalled. It refused to budge. And it now refused to talk to NASA engineers. Instead, it appeared to be rebooting itself, over and over.

Opposite: A model of the Opportunity rover in a simulated Martian environment at JPL.

Project manager Pete Theisinger did not have high hopes for a speedy recovery. "The patient is in intensive care," he said.

Regardless of Spirit's problems, in just over twelve hours, Opportunity was due to land on the opposite side of the planet. A few days earlier, when it fired thrusters in a maneuver to fine-tune its trajectory, Lou D'Amario, the head of MER's navigation team, announced that it had a 99 percent chance of hitting its target, a forty-five-mile-long oval in an area called Meridiani Planum. But then, JPL engineers had also expected Spirit to be communicating normally. So despite the favorable odds, no one was taking success for granted.

The scene had a déjà vu quality. For superstitious reasons, team members wore the same clothes they wore for Spirit's landing. These were not nondescript items but true ceremonial vestments, rooted, if perhaps unconsciously, in the clerical culture of Western engineering. Wayne Lee, the chief engineer for entry descent and landing, wore a shirt that was both patriotic and priestly. It had a blue field with white stars on one side and red and white stripes on the other. NASA administrator Sean O'Keefe wore a totemic polo shirt, the deep rust color of Martian soil, which bore the mission's logo above his heart.

Although the public was enthralled by Spirit's landing, they were impatient with its science. "I'm sure you know by now the Mars rover has stopped sending data," Jay Leno told the *Tonight Show* audience on January 23. "Even the Mars rover is bored with Mars. It's nothing but rocks." Later he speculated on what Opportunity would send back: "Let me guess — maybe pictures of rocks. Some red rocks."

In fact, Opportunity stood a good chance of upstaging Spirit, which, as some hypothesized, had begun misbehaving because it was no longer the center of attention. Opportunity's destination, Meridiani Planum, a flat area about the size of Oklahoma, was geochemically different from Gusev Crater, where Spirit sat. It was also different from any other spot where a robotic probe had previously landed.

What distinguished Meridiani was the presence, determined by the thermal emissions spectrometer on the Mars Global Surveyor, of a bed of gray hematite—an iron-oxide mineral that can form when exposed to liquid water. If Opportunity were to discover sedimentary rock, this finding would buttress the hypothesis that liquid water once existed there. Alternatively, if Opportunity were to find only volcanic rocks, this would strike a blow to the liquid-water theory. Gray hematite can also result from the direct oxidation of hot lava filled with iron.

* * *

After the briefing, I sat at a picnic table outside von Kármán Auditorium, nursing a cup of tea in the frail January light. I thought of how far Opportunity had traveled since I watched it leave the Earth—more than 450 million miles. And of the distance I myself had come since I climbed Mariner Road for Shirley's seminar.

What we know about Mars has also evolved considerably since Mariner 6 and Mariner 7 swept by the planet thirty-five years ago. Not only have four U.S. spacecraft successfully landed there, but orbiters such as the Mars Global Surveyor and Mars Odyssey continue to chart the planet. Mysteries, to be sure, remain; but many have been solved.

I feel this way about my father. When I arrived at JPL, I

harbored an old grudge against him. His coldness, I felt, had devastated my mother and me. I believed what he had wanted us to believe: that he had had an important role on Mariner 69, and that his remoteness, as well as his relentless assertions of masculinity, had to do with his significance in the outside world.

But the evidence did not bear this out. The only thing worse than having your darkest beliefs confirmed, I learned, is having them disproved. I was afraid to dig deeper—fearing he would turn out to be another Duke of Deception, the name that memoirist Geoffrey Wolff gave to his father, a con artist who impersonated various professionals, including an aerospace engineer. The dread intensified when I discovered that Wolff's dad and my father had both worked at Northrop. "Northrop," I scribbled on a three-by-five-inch card, "hires lying fathers."

Yet my father had not lied. Inflated, yes, but not made things up. And with the perspective of thirty-five years, his inflation seems poignant. Of course he had exaggerated his importance at home and oppressed us with his masculinity. In the outside world, because his role on the Mariner mission was not dominant, or archetypically masculine, it could be perceived as subordinate, or archetypically feminine. And this was not a way that a midcentury man would have wanted to be viewed.

Nor can one entirely blame the culture of engineering for his horror at appearing feminine. Such horror churned through popular culture. Who, for example, can forget Jim Backus as the apron-wearing dad whose uxoriousness crippled James Dean's character in the movie *Rebel Without a Cause*? Or the torture visited on an effeminate schoolboy in the film *Tea and Sympathy*?

A year or so after Shirley's workshop, while unpacking

boxes that had been in storage for decades, I stumbled upon an object that I had forgotten. It was a rich-wool, mint-green muffler that my father had knitted. In the late 1970s, Roosevelt Grier, a physically immense, irreproachably manly football player, published a book on knitting, which someone had given my father as a joke. To continue the joke—and tweak his feminist daughter—he made me a scarf. The scarf, however, wasn't rough or lumpy or error-ridden. It was soft and faultless and beautiful—an emblem of a feminine element in him that only this once had found expression.

What had made my father a good knitter also made him a good engineer: attention to detail. This attention is why I can't entirely write off his contribution to Mariner 69. As Des Arthurs had reminded me, the most advanced spacecraft, run by the fanciest software, is only as reliable as its simplest mechanical components. If they fail, it fails.

The challenge was to place my father in the context of a larger story—the story of space exploration at midcentury. His life belied the American faith in boundless expansion. It did not have the forward and upward trajectory of a space mission; it went backward and downward. A golden child, he was the son of a vice president of International Harvester. But the freak accident that killed his father also killed much of his will to prevail. And the Great Depression sapped a good deal of the rest.

Yet in the first few decades of the twentieth century, engineering had been a democratic career—one where men from poor families, such as Frank Malina, or men who were the first in their families to go to college could, through pluck and merit, succeed. It was not nepotistic then; nor is it now. "At JPL, it doesn't matter what your father could do," Robert Gounley, a JPL engineer, observed. "It matters what you can do."

More than by circumstances, I think, my father was held back by memory—a childhood memory, painted in false colors by shame and guilt. The memory was so powerful that it isolated him from his fellow engineers, and from the upward thrust of their success. He convicted himself of a crime: missing his father's funeral. And too proud to seek absolution through therapy, he stubbornly refused parole.

My mother, too, did not fulfill her promise. But in her willingness to relinquish a career after marriage, she was typical of her generation, not an exception. I felt her disappointment, though. It showed in the way she pushed me, and in the pleasure she took when I earned good grades.

I knew my parents under stress, crushed by illness and uncertainty. But I met different people in their correspondence. During their engagement, when he lived in San Diego and she in New Orleans, they had exchanged hopeful letters. They placed their faith in God, in the future, and, with heartbreaking innocence, in the technological breakthroughs of the 1950s. In one letter, my mother commiserated with my father on his long hours at Convair, pleased, however, that he was feathering their nest. "Offered Mass and Holy Communion for you," she wrote, "so don't worry."

In reply, he described that nest, which now included a small, secondhand General Electric television that he had just purchased, exceeding their meager budget. "Please say I did the right thing," he wrote. In subsequent weeks, he told her all the shows he longed to watch with her, from football games to the "Voice of

My father, circa 1930.

Firestone with Thomas L. Thomas" to "Ina Ray Hatton and her all-girl band."

"As to my birthday present," he wrote in March 1952, "you suggested a topcoat, but really, dearest, I will probably use it so seldom that I would like to make a counter-suggestion. My slide rule is about 20 years old and I could use a new good one. The Keuffel and Esser is my first choice."

Three weeks before their marriage, in April 1952, he brimmed with optimism—a sentiment I had never seen him so openly express. "Hi, Snooks!" his letter began. "Just think only 21 more days." He described Mary Star of the Sea, the Spanish-style church in downtown La Jolla where the wedding would occur. "There are so many things to talk about," he wrote. "But we will have a lifetime, with many happy moments and, I hope, not too many heartaches."

In the last year of his life, I communicated with him daily by postcard. He wrote back in an unsteady hand. We signed our letters, "Love." The phone was used sparingly, because our conversations often went awry. "You'll never finish that doll book," he snapped into the receiver. "You don't have the concentration." And rather than defend myself, I hung up—reduced by his sentence to a terrified twelve-year-old.

Because I had not gone into a technical field, I thought he dismissed what I did. He never said, "Good job." He said, "I would have done this differently," or "Why are you wasting your time on that?" Yet after his death, I found a sad thing among his possessions, a thing that could stand as an emblem of all our misunderstandings. It was a fat, yellowed, dog-eared scrapbook of my articles, which, a stranger told me at his funeral, he had proudly and persistently inflicted on his friends.

Neither in my family's past nor at JPL did I find what I had expected. But as any experimental scientist will tell you, investigations take on a life of their own. And sometimes lead to startling destinations.

* * *

Opportunity was due to alight at 9:05 p.m. Pacific standard time. After 7:30 p.m., when the prelanding vigil began, a huge screen dominated von Kármán Auditorium. On it were moving images from the control room and animations of the spacecraft as it approached the planet. I'd like to say that anticipation kept me on the edge of my seat. But I was distracted by nonengineering *mishegoss* swirling around the lab.

Everybody loves a winner, and in the past three weeks, politicians from Pasadena and beyond had during photo opportunities soaked up luster from the rover team. On January 15, Vice President Dick Cheney visited JPL, a day after President George W. Bush announced a plan for a return to the moon by 2020 and, at some mysterious future date, a human voyage to Mars.

Clearly, in the view of advertisers, a successful space mission sells—no matter how tangentially that mission is related to a product. Mars-themed ads proliferated after the 1997 Pathfinder landing. Among the more outrageous, my favorite was a pitch for Sonicare toothbrushes that showed a tiny spaceship hovering near the Red Planet. "If you think it's hard getting to Mars," the caption went, "try getting to plaque three millimeters below the gumline."

On the night of Opportunity's landing, paparazzi flanked the

NASA administrator Sean O'Keefe; NASA associate administrator for space science Ed Weiler; JPL director Charles Elachi; MER project manager Pete Theisinger; MER flight systems manager Richard Cook; and MER entry, descent, and landing development manager Rob Manning celebrate Opportunity's landing, 2003.

doorway to the visitor's center. In addition to accommodating journalists in von Kármán Auditorium, JPL was in plusher quarters hosting a celebrity bash, to which California governor Arnold Schwarzenegger and former vice president Al Gore had been invited. No celebrity, however, could upstage the actual landing. At the point when all systems were "Go for EDL"—that is, entry, descent, and landing—attention swiveled toward Mars.

Opportunity hit the Martian atmosphere at about twelve thousand miles per hour. It was slowed to around 950 miles per hour by atmospheric friction and was slowed even more by its

JPL director William Pickering, James Van Allen (discoverer of the Van Allen radiation belts), and Wernher von Braun raise a model of Explorer 1 after its successful launch, 1958.

parachute, which opened about two minutes before landing. In the seconds before impact, its airbags inflated. Retrorockets on its upper shell ignited to break its fall. And it disconnected from its parachute. Then it struck the ground and bounced.

It kept bouncing. When it finally stopped, the team whooped and cheered. Theisinger hugged Rob Manning. Trosper hugged Cox. And Lee, in that wacky flag-themed shirt, brandished a broom, indicating, as a NASA TV commentator explained, a "clean sweep." Schwarzenegger and Gore wandered through the control room, mugging, congratulating, and—in a joke that quickly grew stale—promising to stay away from the controls.

Even Spirit took a turn for the better. While the world focused on Opportunity, a team of engineers had found its problem: a software glitch, which affected its so-called flash, or short-term memory. They put a plan in place to treat it.

Speakers at the postlanding press conference had reason to be proud. They held hands in a line, arms raised, like triumphant prizefighters. The formation recalled a famous image of von Braun, Pickering, and physicist James Van Allen, holding aloft a model of Explorer 1, after receiving confirmation that the satellite had achieved orbit.

"The NASA team led by JPL really swept a doubleheader," said Ed Weiler, NASA associate administrator for space science. "We resurrected one rover and saw the birth of another."

Richard Cook was refreshingly modest. Victory had apparently not washed the Polar Lander from his mind. "Landing on Mars is really, really hard," he said. "The line between success and failure is narrow. We've got to stay humble."

Pete Theisinger thanked the team behind the team. "The people clapping are all the people who haven't seen their spouses for six months," he said.

* * *

Soon after the landing, the VIPs poured out of JPL. The lab felt like a lab again. I had pierced the world from which I had been excluded as a child. My hard-won press badge announced that I belonged. And I understood the powerful grip that this world had upon my father—because it had an equally powerful grip on me.

On Mars, if all was going as planned, Opportunity was righting itself. It had plopped down on a side, rather than the base, of its flowerlike lander—a situation that Pathfinder and Spirit, also cushioned by air bags, had been lucky enough to avoid. I pictured it yanking itself up by its bootstraps, then—in a manner typical of tourists everywhere—taking some snapshots of itself in front of its destination.

When the Mars Odyssey orbiter passed overhead, Opportunity would send these snapshots, or "thumbnails," to the orbiter. Then the orbiter would relay them to Earth in what John Callas, a JPL engineer commenting on the mission for NASA television, referred to as one big information "dump." Slightly more than an hour into the new day, the thumbnails would materialize in the control room.

The crowd in von Kármán Auditorium sighed with relief when a picture of the rover and lander appeared. Opportunity was safe and intact. Next it would send a black-and-white thumbnail of what it saw, the view from its "eyes" to the Martian horizon.

I was not prepared for this picture. Journalists around me gasped. It was not the boring landscape that had made Jay Leno sneer. A strange ridge of dark stones fanned out in the distance, with a light material, like grout, between them. It might have been a giant's Scrabble board or a spidery grid blackened erratically by a child.

"It looks like a tile patio," someone blurted in the control room.

Opportunity's "thumbnail" image of Mars, which startled journalists and scientists alike.

But Callas would not brook such banality. "No human being has ever seen what we're seeing," he emphasized. "This is a first look at a new world."

Steve Squyres was asked to weigh in on the picture. He looked rumpled and sleep-deprived, even more like a rakish rock-

'n'-roll star than he had during Opportunity's launch. Because the Martian day, or Sol, is thirty-seven minutes longer than a day on Earth, the science team was keeping irregular hours — hours that, in the three weeks since Spirit had landed, wreaked havoc on their circadian rhythms.

Squyres gaped at the thumbnail image, blinking. He did not rush to answer.

Two months after the landing, on March 23, the rover science team made a momentous announcement. It had to do with what was found at this outcrop. Not only had Meridiani Planum been covered with water, but there had been a lot of it, and it was salty. They learned this from a pattern of ripples in the sedimentary rock and some traces of chlorine and bromine. Opportunity, they concluded, stood on the bed of a vast ancient sea — a cauldron that might once have contained the molecules of life.

Yet if on the night Opportunity landed, Squyres had a sense of what was ahead, he did not reveal it. He was the consummate cautious scientist. "I will attempt no science analysis because it looks like nothing I've ever seen before in my life," he said.

But he could not squelch his excitement. "That outcrop in the distance is just out of this world," he blurted.

Then he gave voice to what I suspect every scientist, technician, kid, backyard astronomer, or curious insomniac tuned to NASA TV was thinking — a thought that I'm sure had burned in my father's head as an engineer and has at some point galvanized every human explorer.

"I can't wait to get there," he said.

ACKNOWLEDGMENTS

Like the Cassini mission to Saturn, which was launched in 1997 and arrived at the Ringed Planet in 2004, *Astro Turf* did not reach its destination overnight. Nor did it follow a direct route. Many people who are not featured in the text provided rich and essential background. Todd Barber, Z. Nagin Cox, Randy Herrera, and Vallerie Wagner gave generously of their time and thoughts while they worked on the Galileo mission to Jupiter. I will never forget observing Ganymede through the telescope at Caltech as it executed its orbital dance around Jupiter—thank you, Todd, for making this possible. Nor, speaking of dancing, will I forget Leo Cheng's lunchtime swing dance workshop at JPL. Galileo navigators Lou D'Amario and Dennis Byrnes made orbital mechanics comprehensible; the spacecraft's constraint-defying trajectory, in whose design they played a part, deserves its own book, which I'm sure will someday be written.

Other people I interviewed early in this project—Christopher Hartsough, Amy Ryan, Hector Del Castillo, Kauser Dar, Matt Landano, Bob Mitchell, Neal Ausman, Marcia Segura, Stephanie Wilson, Margaret Kivelson, Homer Joe Stewart, Al Hibbs, Larry Dumas, Bruce Murray, Sylvia Miller, Kari Lewis, Suzanne Smrekar, William J. O'Neill, Greg La Borde, Carol Polanskey, the late William J. Pickering—shaped the book, although I was not able to tell all of their fascinating stories. Donna Shirley not only shared her experiences but put me in touch with Alice

Fairhurst, Susan Foster, Kay Haines, Mike O'Neal, and David Santiago. Vivian Weil introduced me to the work of David F. Noble, whose *A World Without Women: The Christian Clerical Culture of Western Science* contributed significantly to the way I viewed the experiences of women at JPL. Conversations with Ralph Blumenthal, Susan Faludi, Nancy Furlotti, Marsha Ivins, David Hay, Sara Lippincott, Vance Muse, Dava Sobel, Ellen Handler Spitz, Robin Swicord, Margaret Wertheim, and Robert Wilson enriched my thinking, and Douglas Cooper's gift for wordplay shaped a critical aspect of the book. Mary Lamont ensured that no transcriptions were garbled. In addition to knowing a great deal about ICBMs, Kevin Lewis directed my attention to "The Bride Comes to Yellow Sky." Paul Wallich of Echo provided some helpful scientific explanations. Michele Amundsen scoured photo archives for the right images. And Roger Malina's vivid recollections of his father made Frank Malina come alive for me.

In the JPL public information office, thanks to George Alexander, Mary Beth Murrill, Frank O'Donnell, Jane Platt, and Veronica McGregor. In the JPL archives, thanks to John Bluth, Charles Miller, and Russ Castonguay. At Northrop Grumman, thanks to Carol Klammer for assistance in locating the men of Mariner Mars 69.

I benefitted from the editorial guidance of William Whitworth, Laurie Essig, Terri Jentz, Caryn R. Leland, Christine Steiner, and Jere Pfister, the technical guidance of Charles Kohlhase and Robert Gounley, and the historical guidance of Spencer Weart and Jessica Wang. I am also grateful to the Alfred P. Sloan Foundation for its generous support.

Many thanks to the Lord family historians — Mike Lord, Beth Wilson, Gael Lord, and Stina Johnson — for details about my father, as well as to Lillian Patton, for introducing me to the Northrop retirees. Mike Orozco and Bob Reese also provided background material about Northrop. Thanks to Maya Kazan for thematic inspiration, to Krystyna Skalski for the spot-on jacket, and to Patricia Williams for her original look at JPL. Thanks also to Jim Campbell and John and Lynn Lasani for inviting me to the thirtieth anniversary celebration of Mariner Mars 69, which unfortunately took place after this book was written.

I am ever grateful to my steadfast agent, Eric Simonoff, who did not flinch when this project took an unexpected turn. And most especially to George Gibson, a hands-on editor and visionary publisher, who helped coax a finished sculpture from a rough marble slab.

Notes

Introduction
or *The Bride Comes to Yellow Sky*

4 "Let your eyes relax . . .": Shirley's workshop took place March 3–5, 1997.

8 "study of failure": Tony Spear, presentation at "Ethics and Success: Lessons from the Pathfinder Mission Conference," Illinois Institute of Technology, Chicago, Illinois, October 1, 1998.

15 "foreign condition": Stephen Crane, "The Bride Comes to Yellow Sky," in *The Open Boat and Other Stories* (New York: Dover, 1993), p. 88.

15 "a creature allowed . . .": Ibid.

18 Significant source for Arthur Rudolph's involvement with slave labor at Mittelbau-Dora: Michael J. Neufeld, *The Rocket and the Reich: Peenemünde and the Coming of the Ballistic Missile Era* (New York: Free Press, 1995), p. 227.

19 Baron-Cohen and Asperger's syndrome: Reuters, "Were Einstein, Newton Autistic?" May 1, 2003. (See also Oliver Morton, "Think Different?" interview with Simon Baron-Cohen, *Wired*, December 2001, and Geoffrey Cowley, "Girls, Boys and Autism," *Newsweek*, September 8, 2003.)

Mariner Mars 69
or *A Foot Soldier's Story*

22 "Never forget, Son . . .": Donald Reilly cartoon, *New Yorker*, reprinted in Charles P. Boyle, *Space Among Us: Some Effects of Space Research on Society*

(Washington, D.C.: Aerospace Industries Association of America, 1974), p. iii.

24 "Tractors and Buffaloes . . .": *Harvester World* (Chicago: International Harvester, October 1919), p. 18.

24 "Better farming is . . .": Ibid., p. 3.

25 "after a night . . .": Ibid., p. 15.

25 "He leaves a sense of personal loss . . .": Ibid.

25 "widow . . .": Ibid., p. 19.

29 JPL downsizing: Dan Weikel, "JPL Says It Will Privatize 1,600 Jobs," *Los Angeles Times,* January 27, 1997.

29 "the Rock": Interview with Ernie Kling, November 8, 2002. (All Kling quotations are from this interview.)

33 "Father is Home . . .": Paul R. Hanna and Genevieve Anderson Hoyt, *At Home* (Chicago: Scott, Foresman Co., 1956), p. 59.

36 Betty "just wants to hook . . .": *Why Study Science?* (1955). The film is in the collection of Rick Prelinger, who makes it and other 1950s titles available online at *www.archive.org/movies/bytitle.htm.* (All quotations are from the online version of the film.)

38 "Many men have worked . . .": Edward Radlauer, and Ruth Shaw Radlauer, *About Missiles and Men* (Chicago: Melmont, 1959), p. 5.

41 "That Mars is inhabited by beings . . .": Percival Lowell, *Mars and Its Canals* (1906), quoted by Patrick Moore, *Patrick Moore on Mars* (London: Seven Dials, 1998), p. 9.

42 Wernher von Braun, "The Mars Project," originally serialized in the April 24, May 1, and May 8, 1960, issues of *This Week,* quoted and condensed by Frederick I. Ordway III, "A Fictional Trip to Mars—Courtesy Wernher von Braun," *Ad Astra: The Magazine of the National Space Society,* January–February 2000, pp. 40–45.

43 place the spacecraft into a "transfer orbit": Interviews with Charles Kohlhase, January 16, 2001, and February 12, 2002. (All Kohlhase quotations in this chapter are from these interviews.)

46 He showed me photos of tests at JPL: Interview with Des Arthurs, December 5, 2001.

55 Mars looks "like Mars": Robert Leighton, quoted by John Noble Wilford, *Mars Beckons: The Mysteries, the Challenges, the Expectations of Our Next Great Adventure in Space* (New York: Vintage, 1991), p. 61.

58 NASA defrauded: Associated Press, "NASA Records Show Faulty Parts, Fraud and Theft of Moon Rocks," *New York Times*, October 31, 2002.

The Rockets' Red Glare, Part 1
or *The Unlikely Beginnings of the Jet Propulsion Laboratory*

60 "left a successful career as editor . . .": John A. Stormer, *None Dare Call It Treason . . . 25 Years Later* (Florissant, Mo.: Liberty Bell Press, 1992), p. iii.

60 East Coast establishment "kingmakers": Phyllis Schlafly, *A Choice Not an Echo* (Alton, Ill.: Pere Marquette Press, 1964), p. 6.

61 "tax-free purse . . . one Communist mistress": Ibid., p. 16.

63 "Do what thou wilt . . .": Theodore von Kármán with Lee Edson, *The Wind and Beyond: Theodore von Kármán: Pioneer in Aviation, Pathfinder in Space* (Boston: Little, Brown, 1967), p. 257.

64 Rockets . . . crackpot stuff: Ridicule suffered by rocketry pioneer Robert Goddard is detailed in Paul Dickson, *Sputnik: The Shock of the Century* (New York: Walker, 2001), pp. 39–46.

65 [footnote] Rocketry must "be avoided in dignified . . .": Ibid., p. 42.

65 "cold sweats": Interview with Frank Malina by Mary Terrall for Caltech Archive, December 14, 1978.

66 "My dad never talked . . .": Interview with Roger Malina, July 31, 1999. (All Roger Malina quotations are from this interview.)

67 "Forced to leave . . .": Jerry Grey, *Enterprise: The Use of the Shuttle in Our Future Space Programs—the Dreams, the Battles, and the Personalities Involved* (New York: Morrow, 1979), p. 119.

67 "You have to understand how it was . . .": Interviews (via telephone and in-person) with Liljan Wunderman, December 22, 1999, and January 5, 2000. (All Wunderman quotations are from these interviews.)

68 "prewar political discussion group": Clayton R. Koppes, *JPL and the American Space Program: A History of the Jet Propulsion Laboratory* (New Haven: Yale University Press, 1982), p. 30.

68 Sidney Weinbaum, head of Professional Unit 122: Prosecution and conviction of Weinbaum are detailed in Iris Chang, *Thread of the Silkworm* (New York: Basic Books, 1995), pp. 158–59.

69 The show Eisenhower wanted his generals to see: Dennis Piszkiewicz, *Wernher von Braun: The Man Who Sold the Moon* (Westport, Conn.: Praeger, 1999), p. 88. (Piskiewicz says Eisenhower's request for "Man in Space" is an "oft-told" story, though he could find no memo to corroborate it.)

71 "I took religion very seriously . . .": Malina interview with James H. Wilson, June 8, 1973, JPL Archives.

71 "But my father was an agnostic": Ibid.

72 "Kármán asked me why . . .": Malina to parents, December 20, 1936.

72 "We had seminars . . .": Interview with Homer Joe Stewart, December 2, 1999. (All Stewart quotations are from this interview.)

73 "The oxygen hose ignited . . .": Malina to parents, November 1, 1936.

74 "good intelligent friends": Ibid., June 29, 1936.

75 "delightful screwball": Von Kármán, *The Wind and Beyond*, p. 257.

76 Parsons and Foreman . . . "are like inventors . . .": Malina to parents, April 7, 1940.

76 "second father": Roger Malina attributed this perception to his father.

77 The "rocket paper" . . . is "getting longer . . .": Malina to parents, February 20, 1937.

78 "Smith and I received a letter from a parachute jumper . . .": Ibid., May 22, 1938.

79 FBI informants placed Malina and Wunderman at meetings of Professional Unit 122: Memos, Malina FBI file, January 4, 1952.

79 "When I first came to the laboratory . . .": Dorothy Lewis interview with James H. Wilson, op cit., JPL Archives.

80 "Kármán, with a Jewish background . . .": Malina interview with James H. Wilson, op cit.

80 the plane weighed 753 pounds: Plane weights listed and an actual JATO on display in a vitrine in JPL Archives, April 23, 2004.

83 "We now have something . . .": Malina to parents, March 20, 1942.

83 Details of German JATO research: Michael J. Neufeld, *The Rocket and the Reich: Peenemünde and the Coming of the Ballistic Missile Era* (New York: Free Press, 1995), p. 150.

84 "We are getting to be more and more like capitalists . . .": Malina to parents, October 24, 1943.

84 "An accountant went through . . .": Ibid., September 3, 1942.

85 "Babylon the Great, the Mother of Harlots . . .": Description of Babylon in the Book of Revelation (chapter 17, v. 5), quoted by Jon Atack, "Hubbard and the Occult," FACTnet report, http://www.religio.de/atack/occl.html#18.

86 "We were well along . . .": Malina interview with Terrall.

88 "a monstrous place": Von Kármán, *The Wind and Beyond*, p. 279.

88 "Each man was given . . .": Ibid.

89 "exploit . . . chosen rare minds . . .": Joint Chiefs of Staff memo quoted by Christopher Simpson, *Blowback: The First Full Account of America's Recruitment of Nazis, and Its Disastrous Effect on Our Domestic and Foreign Policy* (New York: Weidenfeld and Nicholson, 1988), p. 33.

89 Details of unsanitary conditions at Mittelbau Dora: Michael Neufeld, introduction to Yves Beon, *Planet Dora: A Memoir of the Holocaust and the Birth of the Space Age* (New York: Westview Press, 1997), p. xiv.

89 "Dora's death toll was effectively 6,000 in six months.": Ibid., p. xv.

90 "More people died producing it . . .": Michael Neufeld, *The Rocket and the Reich: Peenemünde and the Coming of the Ballistic Missile Era* (New York: Free Press, 1995), p. 264.

90 Such claims "must be regarded with the greatest skepticism . . .": Ibid., p. 227.

90 "to seek out more qualified detainees": Ibid., p. 228.

90 "American Intelligence officials concealed . . .": Ralph Blumenthal, "Nazi Whitewash in 1940's Charged," *New York Times*, March 11, 1985.

92 a "sensation" among scientists: Ralph Blumenthal, "Drive on Nazi Suspects a Year Later: No U.S. Legal Steps Have Been Taken," *New York Times*, November 23, 1974.

92 "The government closed ranks . . .": Interview with Ralph Blumenthal, August 20, 2004.

92 "rat shack": Piszkiewicz, *Wernher von Braun*, p. 55.

92 "We are living with Army men . . .": Malina to parents, September 29, 1945.

93 "I soon realized that Louis Dunn's attitude . . .": Malina interview with Terrall.

93 "a brilliant man . . .": Lewis interview, with Wilson.

94 "I was getting caught up . . .": Malina interview with Mary Terrall, op. cit.

94 "He is out to win this war . . .": Malina FBI File, November 28, 1942.

96 "passed out Communist literature": Memo, Malina FBI file, November 28, 1942.

96 "Like almost all scientists . . .": Malina to parents, April 2, 1947.

97 "English, very fine . . .": Ibid., July 17, 1948.

97 "As the ceremony was in French . . .": Ibid., March 14, 1949.

97 "looked like a little frog . . .": Ibid., July 8, 1950.

97 "UNESCO is being strongly attacked . . .": Ibid., July 7, 1942.

97 "In the *Washington Times Herald* . . .": Memo, Malina FBI file, November 30, 1951, pp. 3–4.

98 "a Communist front": Ibid., July 27, 1951.

98 Mailing address mix-up: Ibid., August 23, 1949.

98 "reliable" source: Ibid., January 4, 1952.

99 "insufficient documentary evidence . . .": Ibid., August 7, 1952.

99 "authorize prosecution": Ibid., October 10, 1952.

100 "considerable interest in this matter by the Bureau": Ibid., January 12, 1953.

100 "fugitive": Ibid., January 2, 1953.

100 "must be immediately intensified . . .": Memo from FBI director J. Edgar Hoover to FBI office in Los Angeles, Malina FBI file, February 18, 1953.

101 Deported, Tsien went on to found the Chinese ICBM program: Koppes, *JPL and the American Space Program,* p. 31. Also, with the launch of *Shenzhou 5* in 2003, Tsien was celebrated as the father of the Chinese space program. Joe McDonald, "China Scientist Has Career Roots in U.S.," Associated Press, October 15, 2003.

101 "Public employees have been subjected . . .": John Whiteside Parsons, *Freedom Is a Two-Edged Sword* (Las Vegas: Falcon Press, 1989), p. 9.

102 "a 'character,' an eccentric . . .": Memo, Malina FBI file, March 4, 1953.

102 Parsons's "carelessness": Interview with William Pickering, November 10, 1999. (All Pickering quotations are from this interviw.)

103 "This was the great scientific secret . . ." Abraham Polonsky, *A Season of Fear* (New York: Cameron Associates, 1956), p. 21.

103 "a do-gooder . . .": Ibid., p. 17.

103 "had given considerable thought . . .": Memo, Malina FBI file, October 9, 1961.

The Rockets' Red Glare, Part 2
or *Portrait of the Artist*

104 Aerojet valued at four hundred thousand dollars: Memo, Malina FBI file, January 27, 1960.

105 "He certainly does turn out stuff in quantity . . .": Malina to parents, December 12, 1948.

105 "I want to look . . .": Ibid., February 15, 1953.

105 "in making oil paintings of bowls . . .": Interviews with Marjorie Malina, September 24, 25, and 27, 1999. (All Marjorie Malina quotations are from these interviews.)

106 "real stinkers": Malina to parents, February 16, 1954.

106 Dismissal of indictment against Malina: Memo, Malina FBI file, February 16, 1959.

107 a "natural conclusion of Mondrian's . . .": Reg, Gadney, *The London Magazine*, August 1964, p. 40.

107 "History shows that a strongly negative reception . . .": Malina to parents, October 31, 1953.

108 "courteous and affable": Memo, Malina FBI file, November 14, 1956.

109 Bronowski (a regular visitor): Marjorie Malina interview.

109 "vivacious provocateur": Interview with Sandy Koffler, September 24, 1999. (All Koffler quotations are from this interview.)

111 Elves responsible for "Man in Space": Disney's involvement with the Germans is detailed in Howard E. McCurdy, *Space and the American Imagination* (Washington, D.C.: Smithsonian Institution Press, 1997), pp. 41–43.

111 "Man Will Conquer Space Soon": *Collier's*, March 22, 1952.

112 Von Braun's genesis myth: *American Weekly*, July 20, July 27, August 3, 1958, quoted by Piszkiewicz, *Wernher von Braun*, pp. 126–27.

112 International protests of *I Aim at the Stars*: Ibid., pp. 129–30.

114 "for the sake of science": Von Kármán, quoted by Malina in a letter to his parents, April 10, 1937.

115 "The Communists' threat to the free world . . .": Von Braun's speech to Elliot Committee, Piszkiewicz, *Wernher von Braun*, p. 124.

116 Malina's international lunar lab: Memo, Malina November 30, 1964.

117 "thoroughly loyal . . .": Memo of Tieman Dippel interrogation, ibid., February 5, 1965.

119 "This story is to be built . . .": Malina, unpublished screen treatment, 1937, p. 1.

119 "highly sexed . . .": Ibid., p. 2.

119 "Institute of Science . . . bought, unconsciously, by money": Ibid., p. 2.

120 "a wealthy aircraft manufacturer": ibid., p. 17.

121 Arthur Rudolph's fall from grace is detailed in a variety of places, including Michael J. Neufeld, introduction to Yves Béon, *Planet Dora: A Memoir of the Holocaust and the Birth of the Space Age* (New York: Westview, 1997), pp. xxv–xxvi.

Gender Parity, Part I
From Science Fiction to Science Fact

125 "biochemist and ecologist . . .": Robert A. Heinlein, "Let There Be Light," collected in *The Man Who Sold the Moon* (New York: New American Library, 1950), p. 11.

125 "Delilah and the Space-Rigger": Robert A. Heinlein, "Delilah and the Space-Rigger," collected in *The Green Hills of Earth* (New York: New American Library, 1951), p. 13.

126 "the Mother Thing": Robert A. Heinlein, *Have Spacesuit — Will Travel* (New York: Ace Books, 1958), p. 150.

126 "reserving 110 pounds . . .": Von Braun, quoted by Howard E. McCurdy, *Space and the American Imagination* (Washington, D.C.: Smithsonian Institution Press, 1997), p. 224.

126 Heinlein's ruffling of Cronkite and promotion of Fleming: Detailed in Leon Stover, *Robert Heinlein* (Boston: Twayne, 1987), p. 133.

128 "suitable job": Interview with Betty Jane Nolan, January 15, 2004.

130 "are remarkably suited": Alfred Bester, *The Life and Death of a Satellite: A Biography of the Men and Machines at War with Space* (Boston: Little, Brown, 1966), p. 74.

130 "Engineers dare not . . .": Ibid., p. 74.

130 "the conventional wisdom . . .": Interview with Bruce Murray, May 15, 2000. (All Murray quotations are from this interview.)

130 Burbridge turned down Annie Jump Cannon Award: Margaret W. Rossiter, *Women Scientists in America: Before Affirmative Action, 1940–1972* (Baltimore and London: Johns Hopkins University Press, 1995), p. 353.

131 Harvard physician Edward H. Clark: David F. Noble, *A World Without Women: The Christian Clerical Culture of Western Science* (New York: Oxford University Press, 1992), p. 275.

132 "Misogynist Dinner of the American Chemical Society": Ibid., pp. 277–78.

132 Mary Somerville and the Royal Society: Ibid., p. 280.

132 "For many years . . .": Rossiter, *Women Scientists in America*, p. xvi.

133 "When the call went out . . .": Ibid., p. 14.

133 Ellen S. Richard at MIT: Noble, *A World Without Women*, p. 270.

133 "sons and daughters": Eaton, quoted in ibid., p. 266.

133 "that it would be just . . . exciting the militant suffragettes": Scherer, quoted in Judith R. Goodstein, *Millikan's School: A History of the California Institute of Technology* (New York: W. W. Norton, 1991), p. 31.

134 "A conference was called . . .": The *Corporal/Sergeant Story*, copy in JPL Archives.

135 "Section 23": record in ibid.

135 "a shapely craft, 5'6" in height . . .": A campaign for Miss Guided Missile, quoted by Julie Reiz in a presentation at JPL on women at JPL, September 30, 1997.

135 Queen of Outer Space: *The Lab-Oratory*, February 1959, p. 1.

136 "have to get grimy . . .": Interview with Matt Landano, July 11, 1996. (All Landano quotations are from this interview.)

136 "It was really weird as a fresh-out . . .": Interviews with Jan Berkeley, October 8, 1996, and December 12, 2001. (All Berkeley quotations are from these interviews.)

139 "anomalous scattering . . .": Interview with Marcia Neugebauer, June 20, 2000.

140 "rooting for the Army's Jupiter missile . . .": Marcia Neugebauer, "Pioneers of Space Physics: A Career in the Solar Wind," *Journal of Geophysical Research* 102, no. A12 (1997): 26,887.

141 "the center of a three-sided battle": Ibid., p. 26,893.

141 "You don't earn enough . . .": Neugebauer interview.

141 "the Caltech medical staff . . .": Ibid.

142 "fast or hot ions": Neugebauer, "Pioneers of Space Physics," p. 26,892.

143 Terhune's background: Interview with Charles H. Terhune, May 12, 2000. (Terhune self-consciously delivered a *Cliff's Notes* version of his recruitment efforts: "We sent people out to hire lady engineers, and they hired black lady engineers as well.")

143 "women knew better . . .": Interview with Susan Foster, January 21, 1997. (All Foster quotations are from this interview.)

143 Creation of ACW: Interview with Kay Haines, May 11, 2000.

144 "full-blown hippie days" and "They didn't know enough . . .": Interview with Eric Nelson and Elyssa Nelson, October 25, 1996.

145 "learned all the passwords": Interviews with Donna Shirley, July 30, 1996, March 3, 1997. (Unless otherwise indicated, Shirley's quotations are from these interviews.)

145 "partnership model": Riane Eisler, *The Chalice and the Blade: Our History, Our Future* (San Francisco: Harper and Row, 1988), p. xx.

145 "dominator model": Ibid.

145 "lethal power . . . exploitation of their persons": Ibid., p. 49.

145 "linkage": Ibid., p. xvii.

145 "The underlying problem . . .": Ibid., p. xviii.

146 "intervene . . . cultural evolution": Ibid., p. xxi.

146 Shirley management structure: Shirley, "Managing Creativity."

151 "who could express themselves . . .": Interview with Al Hibbs, December 13, 1996.

151 "Donna is way up . . .": Interview with Brian Muirhead, March 3, 1997.

151 "was not a cause for celebration": Brian K. Muirhead and William L. Simon, *High Velocity Leadership: The Mars Pathfinder Approach to Faster, Better, Cheaper* (New York: HarperCollins, 1999), p. 95.

151 "shoo-in": Ibid., p. 94.

151 "overcome a personality conflict . . .": Ibid.

151 "Our best hope": Ibid.

154 "It was like a death": Interviews with Sarah Gavit, August 5, 1999, and April 21, 2000. (All Gavit quotations are from these interviews.)

155 "were not adequately tested": *Mars Program Independent Assessment Team Summary Report*, March 14, 2000, p. 3.

155 "were not ready for launch": Ibid.

155 "inexperienced": Ibid., p. 4.

155 "Fly as you test . . .": Gavit explained this concept to me in March 2000. ("DS 2 had an inadequate test program that deviated significantly from the proven practice of 'test-as-you-fly, fly-as-you-test,'" *Mars Program Independent Assessment Team Summary Report*, p. 3.)

156 "The New Millennium Program . . .": Interview with Kane Casani, November 14, 2000.

No Lost Opportunity
or *The Launch of MER-B*

163 sixteen hundred known problems: John Kelly and Todd Halverson, "Seven Fixes Needed for Return to Flight," *Florida Today*, June 22, 2003, p. 1. (Part one of a seven-part series that ran daily from June 22 to June 28, 2003.)

164 *Top Gun* reversed some antimilitary sentiment: Interview with Philip Strub, May 20, 1998.

165 Measurements of JPL visitor's center: George Alexander, letter to Harrison Price in *Charrette Briefing Materials: Jet Propulsion Laboratory Visitor Center, April 16 and 17, 1997* (San Pedro, Calif.: Harrison Price Company, 1997).

165 "a magic carpet": Interview with George Alexander, March 27, 1998.

166 "Let Pepsi . . .": Interview with Mike O'Neal, May 30, 1998.

166 NASA brain drain: Scott, Gold, "NASA Trying to Reverse Brain Drain: After a Decade of Difficulty Attracting Recruits, Agency Aims to Hire Fresh Talent," *Los Angeles Times*, February 11, 2003.

172 Gene Krantz's vests: Reported by Andrew Chaikin, *A Man on the Moon: The Voyages of the Apollo Astronauts* (New York: Penguin, 1995), p. 190.

173 "On June 17, the MER-B rover . . .": NASA press release, June 18, 2003.

173 "sound-suppression system": Kris Walsh, NASA press conference, Cape Canaveral, June 27, 2003.

174 "tantalizing clues": Steve Squyres, NASA press conference, Cape Canaveral, June 27, 2003.

175 "look good projected . . .": Andrew Bridges, "Twin Mars Rovers to Take Photographs of Red Planet," Associated Press, December 30, 2003.

179 "Place a mark . . .": Interview with Charles Kohlhase, July 7, 2003.

183 "What is called scientific . . .": Evelyn Fox Keller, *Reflections on Gender and Science* (New Haven and London: Yale University Press, 1985), p. 92.

Gender Parity, Part 2
From Science Fact to Science Fiction

185 "potentials": Ursula K. Le Guin, *The Left Hand of Darkness* (New York: Ace Books, 1969), p. 94.

185 "perverts": Ibid., p. 36.

185 "manwoman": Ibid., p. 95.

188 "sexual perversion": Executive Order 10450, quoted by Alan Berube, *Coming Out Under Fire: The History of Gay Men and Women in World War Two* (New York: Free Press, 1990), p. 269.

190 "For every obvious homosexual . . .": *Life,* June 26, 1964, p. 66, quoted by Lee Edelman, "Tea Rooms and Sympathy, or, The Epistemology of the Water Closet," in Andrew Parker et al., *Nationalisms and Sexualities* (New York and London: Routledge, 1992), pp. 263–84. Edelman's article deals with the 1964 arrest of Walter Jenkins, President Lyndon Johnson's chief of staff, for soliciting sex in a YMCA men's room. Lee also discusses the cold-war-era conflation of communism with homosexuality.

190 "Do the homosexuals . . .": *Life,* June 26, 1964, p. 76.

190 "treasonous desire . . .": Laurie Essig, *Queer in Russia: A Story of Sex, Self, and the Other* (Durham, N.C., and London: Duke University Press, 1999), p. 5.

190 "In the sixties . . .": Interviews with Gordon Wood, March 5, 2000, and Gordon Wood and Glenn Cunningham, April 13, 2000. (All Wood quotations are from these interviews.)

191 policy on security clearances: Barbara Spector, "Security Clearance Delays Hamper Gays' Careers," *Scientist,* March 1992, p. 9. See also *Security Clearances for Gay Men and Women* (an unbylined publication) (San Francisco: National Gay Rights Advocates, 1987).

193 "Homophobia in the Sciences": Interview with Cate Heneghan and Annmarie Eldering, December 29, 2000. (Details of the group's early political action are from this interview, as well as from an interview with Randy Herrera, March 4, 2002.)

193 "This subject is not . . .": JPL interoffice memorandum, collection of Cate Heneghan, January 18, 1990. (Also, Heneghan's response to the decision, a memorandum to Robert E. Sutherland, JPL Human Resources Division, January 25, 1990.)

193 high-tech gays: Gary Libman, "Scientists Confront Homophobia in Their Ranks," *Los Angeles Times*, February 2, 1990. (See also Gary Libman, "A New Acceptance: Gay Support Groups Are Beginning to Pay Off in the Workplace," *Los Angeles Times*, July 19, 1990.)

195 "following the water": William J. Broad, "The Hunt for Clues to a Planet's Watery Past," *New York Times*, May 27, 2003.

196 "Is it fair . . .": Arthur C. Clarke and Gentry Lee, *The Garden of Rama* (New York: Bantam, 1991), p. 14.

197 "more comfortable . . .": Ibid., p. 40.

197 O'Toole's devout Catholicism: Ibid., p. 63–66.

199 "At the university . . .": Interviews with Gentry Lee, March 5, 2001, and January 24, 2004. (All Lee quotations are from these interviews.)

200 "Western cultures . . .": Interview with Z. Nagin Cox, October 23, 2001.

203 "Science fiction is metaphor . . .": Le Guin, *The Left Hand of Darkness*, introduction (no page numbers in introduction).

203 "could not think of anyone better": Arthur C. Clarke, Science Fiction Museum and Hall of Fame press release, February 11, 2004.

Bouncing Toward Meridiani
or *Postcards from the New World*

206 "The patient is in intensive care:" Pete Theisinger, quoted in press conference, January 22, 2004.

208 Geoffrey Wolff's dad at Northrop: Geoffrey Wolff, *The Duke of Deception: Memories of My Father* (New York: Random House, 1979), p. 61.

210 "At JPL, it doesn't matter . . .": Interview with Robert Gounley, July 31, 1996.

210 "Offered Mass": Mary Pfister to Charles Lord, October 15, 1951.

210 "Please say": Lord to Pfister, November 20, 1951.

212 "As to my birthday present . . .": Ibid., March 2, 1952.

212 "Hi, Snooks! . . .": Ibid., April 10, 1952.

213 "If you think it's hard . . .": Ad for Sonicare toothbrushes, *Newsweek*, November 4, 1997.

Bibliography

Adams, James L. *Flying Buttresses, Entropy, and O-Rings: The World of an Engineer.* Cambridge, Mass.: Harvard University Press, 1991.

Baron-Cohen, Simon. *The Essential Difference: The Truth About the Male and Female Brain.* New York: Basic Books, 2003.

Beers, David. *Blue Sky Dream: A Memoir of America's Fall from Grace.* New York: Doubleday, 1996.

Béon, Yves. *Planet Dora: A Memoir of the Holocaust and the Birth of the Space Age.* New York: Westview, 1997.

Berube, Allan. *Coming Out Under Fire: The History of Gay Men and Women in World War Two.* New York: Free Press, 1990.

Bester, Alfred. *The Life and Death of a Satellite: A Biography of the Men and Machines at War with Space.* Boston: Little, Brown, 1966.

Bilstein, Roger E. *Orders of Magnitude: A History of the NACA and NASA, 1915–1990.* Washington, D.C.: NASA, 1989.

Boyce, Joseph M. *The Smithsonian Book of Mars.* Washington, D.C.: Smithsonian Institution Press, 2002.

Boyle, Charles P. *Space Among Us: Some Effects of Space Research on Society.* Washington, D.C.: Aerospace Industries Association of America, 1974.

Braun, Wernher von, and Frederick I. Ordway III. *History of Rocketry and Space Travel.* New York: Thomas Y. Crowell, 1969.

Brockman, John. *The Third Culture: Beyond the Scientific Revolution.* New York: Touchstone, 1996.

Burrows, William E. *Exploring Space: Voyages in the Solar System and Beyond.* New York: Random House, 1990.

———. *This New Ocean: The Story of the First Space Age*. New York: Random House, 1998.

Carter, John. *Sex and Rockets: The Occult World of Jack Parsons*. Venice, Calif.: Feral House, 1999.

Chaikin, Andrew. *A Man on the Moon: The Voyages of the Apollo Astronauts*. New York: Penguin, 1995.

Chang, Iris. *Thread of the Silkworm*. New York: Basic Books, 1995.

Clarke, Arthur C. *Rendezvous with Rama*. New York: Harcourt Brace, 1973.

Clarke, Arthur C., and Gentry Lee. *Cradle*. New York: Warner Books, 1988.

———. *The Garden of Rama*. New York: Bantam, 1991.

———. *Rama Revealed*. New York: Bantam, 1995.

Collins, Martin J. *Cold War Laboratory: RAND, the Air Force, and the American State, 1945–1950*. Washington, D.C.: Smithsonian Institution, 2002.

Cooper, Henry S. F., Jr. *The Evening Star: Venus Observed*. New York: Farrar Straus Giroux, 1993.

———. *Imaging Saturn: The Voyager Flights to Saturn*. New York: Holt, Rinehart and Winston, 1981.

Cowley, Geoffrey. "Autism's Gender Gap." *Newsweek*, September 8, 2003, pp. 42–50.

Crane, Stephen. *The Open Boat and Other Stories*. New York: Dover, 1993.

Crowley, Aleister. *Diary of a Drug Fiend*. York Beach, Maine: Samuel Weiser, 1997.

Dethloff, Henry. *Suddenly, Tomorrow Came . . .* Washington, D.C.: NASA, 1993.

Devorkin, David H. *Science with a Vengeance: How the Military Created the US Space Sciences After World War II*. New York: Springer, 1992.

Dick, Philip K. *The Three Stigmata of Palmer Eldrich*. New York: Vintage, 1965.

Dickson, Paul. *Sputnik: The Shock of the Century*. New York: Walker, 2001.

Didion, Joan. *Where I Was From*. New York: Alfred A. Knopf, 2003.

Edelman, Lee. "Tearooms and Sympathy, or, The Epistemology of the Water Closet." In *Nationalisms and Sexualities*, edtied by Andrew Parker, Mary Russo, Doris Sommer, and Patricia Yaeger, pp. 263–84. New York and London: Routledge, 1992.

Eisler, Riane. *The Chalice and the Blade: Our History, Our Future*. San Francisco: Harper and Row, 1988.

Essig, Laurie. *Queer in Russia: A Story of Sex, Self, and the Other*. Durham, N.C., and London: Duke University Press, 1999.

Fairhurst, Alice M., and Lisa L. Fairhurst. *Effective Teaching, Effective Learning*. Palo Alto, Calif.: Davies-Black, 1995.

Fallaci, Orianna. *If the Sun Dies: A Personal Exploration of the World of Tomorrow as It Is Being Entered by the Astronauts and Scientists in the United States Space Program*. New York: Atheneum, 1966.

Faludi, Susan. "Grounded." *Doubletake*, Fall 1995, pp. 28–32.

———. *Stiffed: The Beytrayal of the American Man*. New York: William Morrow, 1999.

Feynman, Richard P. *QED: The Strange Theory of Light and Matter*. Princeton, N.J.: Princeton University Press, 1985.

———. *"Surely You're Joking, Mr. Feynman!" Adventures of a Curious Character*. New York: Bantam, 1986.

———. *"What Do You Care What Other People Think?" Further Adventures of a Curious Character*. New York: Bantam, 1989.

Florman, Samuel C. *The Existential Pleasures of Engineering*. New York: St. Martins Griffin, 1994.

Gleick, James. *Chaos: Making a New Science*. New York: Penguin, 1987.

———. *Genius: The Life and Science of Richard Feynman*. New York: Vintage, 1992.

Godwin, Robert, ed. *Mars: The NASA Mission Reports*. Burlington, Ontario, Canada: Apogee Books, 2000.

Goodstein, Judith R. *Milliken's School: A History of the California Institute of Technology*. New York: W. W. Norton, 1991.

Gorn, Michael. *The Universal Man: Theodore von Kármán's Life in Aeronautics*. Washington, D.C.: Smithsonian Institution, 1992.

Grey, Jerry. *Enterprise: The Use of the Shuttle in Our Future Space Programs — the Dreams, the Battles, and the Personalities Involved.* New York: William Morrow, 1979.

Haley, Andrew G. *Rocketry and Space Exploration.* New York: D. Van Nostrand, 1958.

Hall, R. Cargill. *Lunar Impact: A History of Project Ranger.* Washington, D.C.: Scientific and Technical Information Office, NASA, 1977.

————, ed. *History of Rocketry and Astronautics: ASS History Series.* Volume 7, part 1. San Diego, Calif.: American Astronautical Society, 1986.

Hanna, Paul R., and Genevieve Anderson Hoyt. *At Home.* Chicago: Scott, Foresman, 1956.

Harding, Sandra. *Is Science Multicultural?* Bloomington and Indianapolis: Indiana University Press, 1998.

————. *The Science Question in Feminism.* Ithaca and London: Cornell University Press, 1986.

Harding, Sandra, and Jean F. Barr, eds. *Sex and Scientific Inquiry.* Chicago and London: University of Chicago Press, 1987.

Heinlein, Robert A. *The Green Hills of Earth.* New York: New American Library, 1951.

————. *Have Spacesuit — Will Travel.* New York: Ace Books, 1958.

————. *The Man Who Sold the Moon.* New York: New American Library, 1951.

————. *Rocket Ship Galileo.* New York: Ace Books, 1947.

Henriksen, Margot A. *Dr. Strangelove's America: Society and Culture in the Atomic Age.* Berkeley: University of California Press, 1997.

Hickam, Homer. *Rocket Boys: A Memoir.* New York: Delacourt, 1998.

Hodges, Andrew. *Alan Turing: The Enigma.* New York: Walker, 2000.

Hoover, J. Edgar. *Masters of Deceit: The Story of Communism in America and How to Fight It.* New York: Pocket Books, 1959.

Hunley, J. D., ed. *The Birth of NASA: The Diary of T. Keith Glennan.* Washington, D.C.: NASA, 1993.

Hunt, Linda. *Secret Agenda: The United States Government, Nazi Scientists, and Project Paperclip, 1945 to 1990*. New York: St. Martin's Press, 1991.

Jaynes, Julian. *The Origins of Consciousness in the Breakdown of the Bicameral Mind*. Boston: Houghton Mifflin, 1976.

Johnson, David K. *The Lavender Scare: The Cold War Persecution of Gays and Lesbians in the Federal Government*. Chicago and London: University of Chicago Press, 2004.

Kármán, Theodore von, and Maurice A. Biot. *Mathematical Methods in Engineering: An Introduction to the Mathematical Treatment of Engineering Problems*. New York: McGraw-Hill, 1940.

Kármán, Theodore von, with Lee Edson. *The Wind and Beyond: Theodore von Kármán: Pioneer in Aviation and Pathfinder in Space*. Boston: Little, Brown, 1967.

Keller, Evelyn Fox. *Reflections on Gender and Science*. New Haven and London: Yale University Press, 1985.

Kimmel, Michael. *Manhood in America: A Cultural History*. New York: Free Press, 1996.

Kluger, Jeffrey. *Journey Beyond Selene: Remarkable Expeditions Past Our Moon and to the Ends of the Solar System*. New York: Simon and Schuster, 1999.

Kohlhase, Charles, ed. *The Voyager Neptune Travel Guide*. Pasadena, Calif.: NASA/JPL, 1989.

Koppes, Clayton R. *JPL and the American Space Program: A History of the Jet Propulsion Laboratory*. New Haven: Yale University Press, 1982.

Kuhn, Thomas S. *The Copernican Revolution: Planetary Astronomy in the Development of Western Thought*. Cambridge, Mass., and London: Harvard University Press, 1952.

———. *The Structure of Scientific Revolutions*. Chicago and London: University of Chicago Press, 1962.

Lasby, Clarence G. *Project Paperclip: German Scientists and the Cold War*. New York: Atheneum, 1971.

Le Guin, Ursula K. *The Left Hand of Darkness*. New York: Ace Books, 1969.

MacDougall, Walter A. . . . *The Heavens and the Earth: A Political History of the Space Age.* New York: Basic Books, 1985.

Mailer, Norman. *Of a Fire on the Moon.* Boston: Little, Brown, 1969.

Malina, Frank J., ed. "America's First Long-range Missile and Space Exploration Programme." *Spaceflight* 15, no. 12 (1973).

———. *Applied Sciences Research and Utilization of Lunar Resources: Proceedings of the Fourth Lunar International Laboratory (LIL) Symposium.* New York: Pergamon Press, 1970.

———. *Kinetic Art: Theory and Practice.* New York: Dover, 1974.

McAleer, Neil. *The Authorised Biography of Arthur C. Clarke.* London: Victor Gollancz, 1992.

McCurdy, Howard E. *Space and the American Imagination.* Washington, D.C.: Smithsonian Institution Press, 1997.

Mitchell, Edgar, with Dwight Williams. *The Way of the Explorer: An Apollo Astronaut's Journey Through the Material and Mystical Worlds.* New York: G. P. Putnam's, 1996.

Moore, Patrick. *Patrick Moore on Mars.* London: Seven Dials, 1998.

Morton, Oliver. *Mapping Mars.* New York: Picador, 2002.

———. "Think Different." *Wired*, December 2001.

Muirhead, Brian K., and William L. Simon. *High Velocity Leadership: The Mars Pathfinder Approach to Faster, Better, Cheaper.* New York: HarperCollins, 1999.

Murray, Bruce. *Journey into Space: The First Thirty Years of Space Exploration.* New York: Norton, 1989.

Murray, Charles, and Catherine Bly Cox. *Apollo: The Race to the Moon.* New York: Touchstone, 1989.

Neufeld, Michael J. *The Rocket and the Reich: Peenemünde and the Coming of the Ballistic Missile Era.* New York: Free Press, 1995.

Newton, Clarke. *The Aerospace Age Dictionary.* New York: Franklin Watts, 1965.

Newton, Verne W. *The Cambridge Spies: The Untold Story of Maclean, Philby, and Burgess in America.* Lanham, Md.: Madison Books, 1991.

Noble, David F. *The Religion of Technology: The Divinity of Man and the Spirit of Invention.* New York: Penguin, 1997.

————. *A World Without Women: The Christian Clerical Culture of Western Science.* New York: Oxford University Press, 1992.

O'Neill, Gerald K. *The High Frontier: Human Colonies in Space.* Princeton, N.J.: Space Studies Institute Press, 1989.

Ordway, Frederick I., III. "A Fictional Trip to Mars—Courtesy Wernher von Braun." *Ad Astra: The Magazine of the National Space Society,* January–February 2000, pp. 40–45.

Parsons, John Whiteside. *Freedom Is a Two-Edged Sword.* Las Vegas: Falcon Press, 1989.

Pauls, Michael, and Dana Facaros. *The Traveler's Guide to Mars.* London: Cadogan, 1997.

Penley, Constance. *NASA/TREK: Popular Science and Sex in America.* New York: Verso, 1997.

Piszkiewicz, Dennis. *The Nazi Rocketeers: Dreams of Space and Crimes of War.* Westport, Ct.: Praeger, 1995.

————. *Wernher von Braun: The Man Who Sold the Moon.* Westport, Ct.: Praeger, 1998.

Polonsky, Abraham. *A Season of Fear.* New York: Cameron Associates, 1956.

Radlauer, Edward, and Ruth Shaw Radlauer. *About Missiles and Men.* Chicago: Melmont, 1959.

Rhodes, Richard. *The Making of the Atomic Bomb.* New York: Touchstone, 1988.

Rossiter, Margaret W. *Women Scientists in America: Before Affirmative Action, 1940–1972.* Baltimore and London: Johns Hopkins University Press, 1995.

————. *Women Scientists in America: Struggles and Strategies to 1940.* Baltimore and London: Johns Hopkins University Press, 1982.

Roszak, Theodore. *The Gendered Atom: Reflections on the Sexual Psychology of Science.* Berkeley, Calif.: Conari Press, 1999.

Russell, Bertrand. *The Impact of Science on Society*. London: George Allen and Unwin, 1952.

Schlafy, Phyllis. *A Choice Not An Echo*. Alton, Ill.: Pere Marquette Press, 1964.

Shirley, Donna. "Managing Creativity." Unpublished book, 1997.

Shirley, Donna, with Danielle Morton. *Managing Martians*. New York: Broadway Books, 1998.

Silberman, Steve. "The Geek Syndrome." *Wired*, December 2001.

Simpson, Christopher. *Blowback: The First Full Account of America's Recruitment of Nazis, and Its Disastrous Effect on Our Domestic and Foreign Policy*. New York: Weidenfeld and Nicholson, 1988.

Snow, C. P. *The Two Cultures*. Cambridge: Cambridge University Press, 1959.

Stormer, John A. *None Dare Call It Treason . . . 25 Years Later*. Florissant, Mo.: Liberty Bell Press, 1992.

Stover, Leon. *Robert Heinlein*. Boston: Twayne, 1987.

Sutin, Lawrence. *Divine Invasions: A Life of Philip K. Dick*. New York: Harmony Books, 1989.

Tanenhaus, Sam. *Whittaker Chambers: A Biography*. New York: Modern Library, 1998.

Thomas, Shirley. *Men of Space: Profiles of the Leaders in Space Research, Development, and Exploration*. Philadelphia and New York: Chilton Books, 1962.

Vincenti, Walter G. *What Engineers Know and How They Know It: Analytical Studies from Aeronautical History*. Baltimore and London: Johns Hopkins University Press, 1990.

Waldie, D. J. *Holyland: A Suburban Memoir*. New York: W. W. Norton, 1996.

Wang, Jessica. *American Science in an Age of Anxiety: Scientists, Anticommunism, and the Cold War*. Chapel Hill and London: University of North Carolina Press, 1999.

Wertheim, Margaret. *Pythagoras' Trousers: God, Physics, and the Gender Wars*. New York: Norton, 1997.

Wilford, John Noble. *Mars Beckons: The Mysteries, the Challenges, the Expectations of Our Next Great Adventure in Space*. New York: Vintage, 1991.

Wolfe, Tom. *The Right Stuff*. New York: Farrar Straus Giroux, 1979.

Wolff, Geoffrey. *The Duke of Deception: Memories of My Father*. New York: Random House, 1979.

Index